Breakthrough Technology Project Management

Second Edition

D1468780

E-Business Solutions

Bennet P. Lientz and Kathryn P. Rea
Series Editors

The list of titles in this series includes:

Start Right in E-Business
Dynamic E-Business Implementation Management
Breakthrough Technology Project Management, 2nd ed.
Grow Your E-Business for Success

Breakthrough Technology Project Management

Second Edition

Bennet P. Lientz

John E. Anderson School of Management
University of California, Los Angeles

Kathryn P. Rea

The Consulting Edge, Inc.
Beverly Hills, California

San Diego   San Francisco   New York   Boston   London   Sydney   Tokyo

Butterworth-Heinemann is an imprint of Elsevier Science.

 This book is printed on acid-free paper.

Library of Congress Cataloging-in-Publication Data
A catalog record for this book is available from the Library of Congress.

ISBN 0-12-449968-6

The publisher offers special discounts on bulk orders of this book.
For information, please contact:
Manager of Special Sales
Elsevier Science
200 Wheeler Road
Burlington, MA 01803
Tel: 781-313-4700
Fax: 781-313-4802

For information on all Butterworth-Heinemann publications available, contact our World Wide Web homepage at http://www.bh.com

10 9 8 7 6 5

Printed in the United States of America

Contents

Part I
Improving the Project Management Process

Chapter 1
Introduction

Chapter 2

Developing Your Project Management Process and Strategy

Chapter 3

Managing Multiple Projects and the Project Slate

Part II
Developing Your Project Plans

Chapter 4
The Project Concept

Chapter 5
The Right Project Leader

Chapter 6

Building the Project Team

Chapter 7
Developing the Project Plan

Part III
Managing Projects

Chapter 8
Effective Project Tracking and Coordination

Chapter 9

Software Development

Chapter 12
Technology Projects

Part IV
How to Successfully Address Project Issues

Chapter 13
Business Issues

Chapter 14
Human Resource Issues

Chapter 15

Management Issues

Chapter 16

Technical Issues

Chapter 17
Vendor and Consultant Issues

Chapter 18

How to Implement Improved Project Management

Preface

INFORMATION TECHNOLOGY (IT) PROJECTS FAIL TOO OFTEN

Studies repeatedly point out that 30 to 45% of systems projects fail prior to completion. Over half of all systems projects overrun their budgets and schedules by 200% or more. Combined costs of failure and overruns total in the hundreds of billions of dollars. Failures and problems continue, despite improved tools and technology. Data also indicate that the failed projects were viewed as critically important by management.

FAILURE STATISTICS ARE STAGGERING

The results of several surveys were published in *Computerworld*, a leading systems magazine. Here are some of their results:

- Failed systems projects cost more than $100 billion per year.
- One of every two projects overruns its budget by 180% or more.
- A survey of what was missing in the project management process indicated the following:
  Project office—42%
  Integrated methods—41%
  Training and mentoring—38%
  Policies and procedures—35%
  Implementation plans—23%
  Executive support—22%

WHY ARE IT PROJECTS DIFFERENT FROM PROJECTS OF THE PAST?

Many of the methods and techniques of the past are still being used today, even though the technology, methods, management, and entire environment have changed. Some of the differences are listed in the following table.

Factor	Traditional	Modern
Focus	Single project	Multiple projects
Management attention	Critical path	Management; critical path focusing on risk and issues
Staffing/resources	Full time/dedicated	Part-time/full-time mix; shared
Project	Side by side to business	Processes and systems are linked
Staffing	Best people	Average people with energy
Milestones	Assume that they can be reviewed	Reviews must be selective due to time and resources
Project status	Budget versus actual; % complete	Unresolved issues; future tasks with risk and issues
Large projects	Divide by organization	Divide by risk
Small projects	Often not treated as project	Include as projects
Risk	Often treated in a fuzzy way	Treated tangibly through issues
Lessons learned	Each project treated as unique so that lessons learned are not stressed	Major emphasis on lessons learned
Management expectations	Moderate	High
IT projects	Critical to departments	Critical to the enterprise

There is a need to update the project management approach to reflect the modern environment.

Systems and technology implementation and support are complex and involve many elements, necessitating planning and project management. Management for these projects is different from that for some standard projects in industries such as construction and engineering. The projects often involve new technology with which the project team is unfamiliar. The projects include interfaces with existing systems and other, incompatible technologies. Integration is often required. Given that many people treat software development and programming as an art, it is easy to see why systems projects become even more complex. Nor are the requirements for the systems stable. Business, technology, and externally generated changes can arise in the middle of the project. Systems projects require extensive cooperation between business units, information technology, and management.

Therefore, it is not surprising that half of client–server projects fail or that almost as many reengineering projects suffer the same fate. Managers at more than 60% of the firms in one survey thought that they had implemented purchased software packages incorrectly and had achieved little or no benefit. Firms indicate that when a failure occurs, the direct losses can be in the millions of dollars and the total indirect losses in business are often much more (because the firm was depending on the results of the completed projects for revenue or for cost reductions).

E-BUSINESS FAILURE IS SIGNIFICANT

There is no doubt that e-business is a major force and trend for the early 21st century. The benefits of e-business are well known in all of the media. However, many don't want to discuss the dark side—e-business failures. Most of these are not publicized. After all, what would happen to a company's stock price if the failure were widely known? Some of the causes of e-business failure are the following:

- E-business implementation is treated like a traditional project—a bad idea.
- The scope of the e-business effort is defined as IT only. Business process and organizational change are not included.
- There is an inadequate provision for change in direction. The project is inflexible to change.

This book addresses these issues and more. Most chapters include a section with e-business guidelines.

WHY DO MANY TECHNOLOGY PROJECTS FAIL?

Why do so many technology and systems fail? Why don't people learn from their mistakes and those of others? Complexity is part of the answer. Also, people get caught up in their work. They move from one project or piece of work to the next. Although they continue to use many of the same tools, they do not gather or apply lessons learned. Were this not enough, management and the business depend on technology today as never before. Technologies not only must be implemented correctly but also must be integrated. The bars of standards and expectations have been raised.

Failure also occurs because people manage technology projects the same way they manage other projects. However, technology projects are different. The duration of the project can be long. During that time, the technology advances and can affect the project. The requirements of the business can change. Typically, technology projects are not carried out from scratch. The new project must always be integrated into the fabric of the current systems and technology—called the *architecture* in this book. As part of the project, the project team members may have to learn the technology as they go. These characteristics are different from what one encounters when building bridges, launching new products, or undertaking other, more common projects. On the other hand, many of the lessons learned from the project management can be applied to systems and technology projects.

Another reason for failure is that projects are managed singly—like disjoint construction projects. This does not work for technology because (1) the projects are often interdependent, (2) many projects depend on the same technology and resources, and (3) issues that cross many projects are resolved in contradictory or conflicting ways between projects. A fundamental theme of this book is that technology projects must be managed as a whole, not as individual projects.

WHAT ARE THE BENEFITS OF PROJECT SUCCESS?

With all of this talk about failure, why do projects at all? There are many reasons, including complexity, duration of the work, and the need for organization of the work. If you are successful in better managing single and multiple projects, experience and lessons learned point to the following benefits:

- There are greater benefits to the business, because the purpose and scope are set and supported to provide tangible business benefits.
- Risk can be minimized and managed better because all projects are being managed collectively as well as individually.
- Resources are better managed, utilization increases, and critical resources can be spread across multiple projects.
- There are fewer surprises in project work and schedules, allowing more predictability.
- You get more productivity and results from investment in technology and systems.

PURPOSE AND SCOPE OF THE BOOK

The purpose of this book is to answer the following questions:

- How can the overall technology project management process be improved?
- Which systems projects should be given resources and approved for action?
- How can you better manage all systems and technology projects together?
- How can individual projects be better managed and more successful?
- What are specific guidelines for managing different types of projects?

The scope of the book includes these and other technology areas and addresses these questions:

- What projects should be approved?
- How do you formulate and start projects effectively?
- How do you manage single and multiple projects?
- How do you identify, analyze, and address specific project issues?
- How do you communicate effectively with management, team members, staff, and vendors to obtain results?

WHY USE THE TERM *BREAKTHROUGH*?

This book focuses on a project management approach that differs from the single project monolithic methods of the past (which addressed single projects, assumed that there were dedicated resources to the project, and assumed that purpose, scope, and requirements were fixed). We don't think that this world ever existed in IT, but if it did, those days are long gone. Modern project management methods have to succeed in a new world with the following characteristics:

- Technology is changing rapidly.
- Management changes business direction in response to pressures.
- Competition and industry change create more stress.
- There is limited staff to work on projects.
- People who work on projects also must do their normal nonproject work.

The book uses a commonsense approach. It is a breakthrough in that it is different from the normal project management methods. This approach has the following themes:

- You have to manage multiple projects, not single projects.
- You derive benefit by having the project team use modern software tools for project management.
- Team members on the project work in a collaborative way in which they participate in defining and updating their work and working on issues.
- Risk management is a major focus through identifying, addressing, and tracking issues.
- A high-level structure through project templates and standardization supports cumulative improvement in project performance. Standardization supports analysis and management reporting. However, at the detailed level of all projects, there is flexibility to accommodate the individual characteristics of the project.
- Lessons learned are gathered throughout the projects and applied to project templates so that projects are improved.

All of these themes must be self-reinforcing across the life of the projects. This is shown in the following table. Here the rows are some of the major activities involved in a project. The columns are three elements of the focus of this book. Note how the themes are mutually reinforcing across all of the areas.

Area	Templates	Issues/lessons learned	Collaborative work
Getting started	Project template needed to start	Project concept	Concept analysis
Meetings	Updating areas of the template	Subject of the meetings	Each person defines and updates
Presentations	Structure	Format defined	Joint presentations

Area	Templates	Issues/lessons learned	Collaborative work
Getting the people	Standardization for what is needed	Use issues as a way to get people; use lessons learned for retention	Joint tasks where participants can make friends
Vendor management	Templates fit with our schedule	Common database; vendors participate in lessons-learned meetings	Joint tasks between vendor staff and internal staff
Changes to the plan	Fit within template	Manage through issues	Joint definition of the change

THE AUDIENCE AND WHAT YOU WILL GET OUT OF THE BOOK

Who can use this book? Anyone who is involved in any type of systems and technology project in either private or public sectors—organizations large and small. We do mean organizations of all sizes. The methods developed in this book have been applied to companies with 10 to more than 45,000 employees. Although this seems very general, the lessons learned have been applied in a variety of companies and industries. In addition, the materials have been tested and employed in project management, software, and technology classes in business and management, engineering, computer science, library and information science, architecture, public policy, and medical schools. We have taught these techniques to more than 4500 people in more than 20 countries.

You will obtain guidelines that can help you achieve greater success and reduce the risk of problems and failure in projects. These guidelines are lessons learned from more than 100 projects over the past 25 years. Lessons learned are a part of knowledge management, in which a firm attempts to leverage from its experience, expertise, and knowledge for competitive advantage.

This book has specific advantages and features over others:

- The book addresses the "how" in systems and technology project management. Most other books address the "what."
- Many specific examples from different industries are featured, including such industries as banking, insurance, manufacturing, distribution, transportation, government, retailing, medical services, and energy and natural resources.
- The book is written in an easy-to-use style. Many books are very dry. The style here is intended to make the material more interesting and useful.
- More than 300 individual guidelines and lessons learned are provided, along with ways you can employ them to be more effective.
- Projects involving software acquisition, development, operations, maintenance, enhancement, and technology are addressed in depth.

HOW IS THE BOOK ORGANIZED?

The book is divided into four logical parts:

- Part I: What should your project management strategy be and how should you address multiple projects?
- Part II: How do you establish projects and project plans?
- Part III: How do you successfully manage software development (client-server, intranet, data warehousing), operations, maintenance, and enhancement, software package implementation, and technology (including networking) projects?
- Part IV: How do you cope with specific issues related to personnel, management, technology, and vendors/contractors?

Part IV also contains a chapter on implementation of the approach and tips on how to overcome 30 potential points of resistance. Many of these actions can be taken by an individual project leader; others require wider support. More than 70 organizations have implemented these actions with success.

Chapters are generally organized in the same manner.

- *Introduction.* This lays the groundwork for the chapter and discusses what has been attempted in the past.
- *Approach.* This is the core of the chapter, which gives detailed methods.
- *Examples.* One or more examples are included.
- *Guidelines.* These are specific suggestions and lessons learned that have been developed from experience.
- *Action items.* These are steps that you can take after reading the chapter.
- *Summary.*

Throughout the book, examples are presented that employ specific technologies involved in various industries, including retailing, distribution, manufacturing, banking, and transportation. These are not your typical successful case type examples. Often, those are not of the real world. Here we deal with the dysfunctional, struggling, and failing companies as well as successful organizations. As you might guess, you often learn more from failure than success.

WHAT IS NEW IN THE SECOND EDITION?

The second edition has the following features:

- There is an expanded discussion of risk management, which extends over several chapters.
- There is a new chapter on management issues.

- Existing chapters on issues have been expanded.
- There is more in-depth discussion of estimation, budgeting, and tracking.
- Quality assurance is addressed in more depth.
- Most chapters now have specific guidelines on how to apply the material to your e-business projects.
- There is a major new chapter on implementing the techniques in the book and how to overcome potential resistance.
- A new appendix serves as an issues checklist at the start of a project.
- Useful Web sites for project management, lessons learned, and collaborative effort have been included.

To make the material more accessible, we have included a section called "The Magic Cross Reference." We gave the feature this title because we wanted to draw your attention to it. It is useful for finding one or several of the more than 200 guidelines in the book. The term "magic" is tongue in cheek, but reflects the fact that searching through an index is often not the best way to find materials in a book.

About the Authors

Bennet P. Lientz is Professor of Information Systems at the Anderson Graduate School of Management, University of California, Los Angeles (UCLA). Dr. Lientz was previously Associate Professor of Engineering at the University of Southern California and department manager at System Development Corporation, where he was one of the project leaders involved in the development of ARPANET, the precursor of the Internet. He managed administrative systems at UCLA and has managed over 70 projects and served as a consultant to companies and government agencies since the late 1970s.

Dr. Lientz has taught information technology, project management, e-business, and strategic planning for the past 20 years. He has delivered seminars related to these topics to more than 4000 people in Asia, Latin America, Europe, Australia, and North America. He is the author of more than 25 books and 70 articles in information systems, planning, project management, process improvement, and e-business. He has been involved in over 40 e-business projects in 15 industries.

Kathryn P. Rea is president and founder of The Consulting Edge, Inc., which was established in 1984. The firm specializes in e-business, information technology, project management, and financial consulting.

Ms. Rea has managed more than 65 major technology-related projects internationally. She has advised on and carried out projects in government, energy, banking and finance, distribution, trading, retailing, transportation, mining, manufacturing, and utilities. She has successfully directed multinational projects in China, North and South America, Southeast Asia, Europe, and Australia. She has conducted more than 120 seminars around the world. She is the author of eight books and more than 20 articles in various areas of information systems and analysis. She has been an e-business project leader on 10 projects and has been involved in several startup e-business companies.

Part I

Improving the Project Management Process

Chapter 1

Introduction

PROJECT MANAGEMENT CONCEPTS

DEFINITIONS

Let's start by defining some basic concepts related to projects and project management for single projects. A *project* consists of work that is focused on specific purposes within the boundaries of a defined *scope*. Projects can be of any size or type. A purpose for a project can be a narrow goal related to a specific system or technology, or it can be more extensive to include improvements in a business process. The scope of the project defines what is included in and excluded from the project. A *business process* is a set of regularly performed, integrated business tasks that produce specific outputs. Payroll, sales, and planning analysis are examples of business processes. The *project plan* consists of related tasks or activities and milestones together with resources, costs or budget, and schedules. It is assumed that you are familiar with terms such as critical path, GANTT, and PERT charts.

Moving up to multiple projects, the *project slate* is the set of approved projects that will be funded and provided with resources. It is often the case that the slate is approved by a *management steering committee* or group. *Resource management* is the coordination, management, and control of resources across the projects. How to resolve competing demands among projects for the same resources will be called *resource deconflicting*. If you place all of the schedules together, the result will be called the *master schedule*.

Turning to systems concepts, the systems organization will be referred to as *information systems* or IT. Many other terms have been used over the years: information systems (IS), management information systems (MIS), and data processing (DP). The business organization staff members have often been referred to as *users*. This term can be viewed as derogatory so the business organizations will be considered as

3

business units. The employees in the business units will mostly be referred as *business staff* or employees.

Both the systems and business organizations have objectives and strategies. An *objective* here is a directional goal that is timeless. Examples relate to profitability, costs, effectiveness, and efficiency. These are general and have to be supported by strategies. A *strategy* provides the focus for supporting objectives. Projects then support strategies that in turn support objectives. A strategy might be to expand market share. In information systems an objective is to be responsive to the business. A strategy might be to implement an intranet or Internet solution.

The collection and structure of the hardware, software and systems, and network will be referred to as the *architecture*. A key word here is *structure*. How the pieces of technology go together is critical to the architecture and its flexibility and capabilities. A poor architecture, for example, might have either missing pieces or parts that do not integrate or interface with each other.

Before and during a project, problems, situations, and opportunities can arise. These will be referred to as *issues*. Effective issue management is a key attribute of good project managers. Issues have symptoms that are signs of problems as well as the problem or opportunity itself. Examples of issues are how to allocate a scarce personnel resource among projects, how to address a gap in technology, and how to deal with a specific requirement change. In order to address issues, management makes *decisions*. Decisions are implemented or followed up by taking *actions*. Some tasks in projects have *risk* in that there is a significant likelihood that there may be problems or slippage. Here we will treat the reason behind the risk as an issue. Thus, if you resolve the issues behind a risky task, you have reduced or removed the risk. This is a key association in the book and has been found to be an effective means of dealing with the more fuzzy concept of risk.

A goal in project management is to gather lessons from experience and then apply these to current and future projects so as to be more effective. These will be referred to as *lessons learned.* Lessons learned can take many forms: policies, guidelines, procedures, and techniques. In systems, lessons learned include how to work with a specific business process or how to effectively use a software tool.

Figure 1.1 gives a picture of the systems and business situation as outlined in preceding discussion. As you can see, the systems and business components can be mirrored. The difference is that the business processes correspond to the systems on the systems side.

In the diagram, business issues are reflected in systems issues (on the left). The organization interface is shown on the right. In the center, the systems and the architecture (structure of the technology) support the business processes. This diagram will be employed to relate business-oriented topics to those in systems. This figure has two dashed boxes. One represents the traditional project approach of focusing only on systems and the architecture (labeled "Old"). The second, larger one is the more modern view of including the business processes along with systems and architecture (labeled "New").

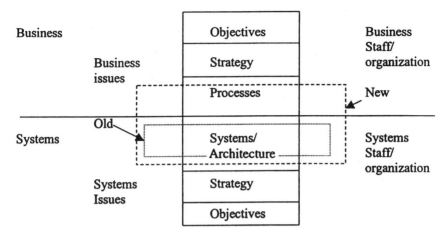

Figure 1.1 Systems and Business Components

DIFFERENCES BETWEEN STANDARD AND IT PROJECTS

In the preface some general differences between standard projects and systems and technology were identified. These can now be made more precise.

- *Purpose.* The goals of a system and technology project are often not as clearly defined as those in engineering or other projects. The goals may not be well defined at the start of the systems project.
- *Scope.* Systems projects sometimes lack clear boundaries. Is the business process within the project? With what systems must the project interface? Moreover, the scope can creep and expand.
- *Parallel work.* While a new system is being created or installed, work can continue on the current system, creating changing requirements. This is not as true in standard projects.
- *Interfacing projects.* Systems projects are more likely to have complex interfaces.
- *Technology dependence.* It seems that only in systems and technology projects do people attempt to use new technology or technology with which they have no or only limited experience, raising the level of risk.
- *Management expectations.* Upper-level managers attend seminars and read about the promise of new technologies. Their expectations can impact the systems project.
- *Cumulative impact.* One project can affect others. The latest project depends on the results of many previous and some current efforts. It is cumulative dependence.
- *Understanding the technology.* Although nonsystems projects can involve technology, it is normally simpler to apply because the technology can be

handled separately. In systems, it is the reverse. The only way modern systems can be successful is by integrating multiple technologies. This requires a deeper and more thorough understanding of the technology.

- *Technology gaps.* Systems and Technology projects are also affected by gaps between the newest technology and the older technologies.

TRENDS IN BUSINESS

Certain business trends have emerged and persisted, affecting systems projects.

- *Accountability.* Business has become more sensitive to accountability and has a greater drive for results—impacting which projects are funded and their schedules.
- *Downsizing/rightsizing.* Information systems units have been hit by this trend as have the business units. The impact is that many organizations have to make up for the decline in staff with automation by putting more pressure on systems projects.
- *Importance of business processes.* Whereas in the past, the focus was on the business organization, attention has shifted to the basic business processes. This new focus has been spurred by the successes of Wal-Mart, Federal Express, and other firms, and it impacts the scope and focus of systems projects.
- *Reengineering and process change.* Changes in business processes are almost always linked to changes in systems and technology.
- *Global competition.* Companies are faced with competing in multiple, diverse markets. This requires flexibility on the part of the technology and systems.
- *Outsourcing/supplier-customer links.* With electronic commerce and electronic data interchange (EDI), firms seek to tighten their links with suppliers and customers. In addition, more functions are being outsourced (in some cases to suppliers and customers).
- *Thirst for information.* Firms desire to get at the information for management, strategic, and tactical/operational purposes. Data warehousing and the spread of database management systems have created new projects that pull information from multiple sources for analysis in support of knowledge management.
- *E-business implementation.* Implementing e-business combines system implementation, process improvement, and organization change.

These factors lead to the creation of more projects that are of increased complexity compared to traditional systems work. Success with the first project in a given area typically leads to more projects in the same areas. For example, a firm that has been successful in reengineering its accounts payable processes is then tempted to move to other accounting areas. Thus, it is not surprising that the number of projects and their interdependencies are growing.

TRENDS IN TECHNOLOGY

Examine some of the key technology trends that have affected project management.

- *Network expansion.* The Internet, local area networks, and wide area networks have made communications and networking more affordable and easier to implement than was previously the case. This means that more companies have implemented networks. The software to handle electronic mail, videoconferencing, and voice communications over networks has improved and become cheaper and less complex to install. Most new systems and technology projects today involve a networking component. Network expansion can be seen in two ways: internal and external. Internally, wide area networks and on-line, client-server applications have fueled network expansion. Externally, the use of electronic commerce, EDI, extranets, and links to suppliers and customers has grown.
- *Declining cost and increased performance of microcomputers.* As a result of this trend, companies have been encouraged to deploy more microcomputers with increased functions to more employees. This makes the systems projects larger because more staff must be trained.
- *Embedded technology.* Technology has become embedded in physical devices as never before. The growth of geographic information systems (GIS) and geographic positioning systems (GPS) are two examples of technology being embedded in vehicles, facilities, and equipment.
- *Cumulative impacts.* With a growing dependence on technology, companies need to manage the cumulative efforts and cross-impacts of the technology.
- *Electronic commerce.* E-commerce is the means of supporting e-business transactions. An increasing number of IT projects deal with e-business and e-commerce.

To many business managers these trends combine to support their beliefs that technology and systems can respond to business trends and needs rapidly and easily. This raises expectations and requirements for systems and technology projects.

TRENDS IN SYSTEMS

The impacts of moving from the raw technology to systems are evident in the following trends:

- *Systems replacement.* Whether dealing with problems or just replacing ancient legacy systems, systems groups are moving to install new software packages. This is not a simple replacement project. The new software offers different features and capabilities.
- *System integration.* There is a trend toward more integration of systems and data. What used to be a series of single projects has been replaced by larger, integrated projects.
- *Improvements in tools.* More tools are available to assist in development and systems management. Object-oriented programming and design tools are emerging in addition to improvements in database and fourth-generation languages as well as data warehousing tools. The approach of using Java and other similar tools for intranets make systems management easier.
- *Stability of vendor software.* As in the old days when there was one dominant hardware vendor, the emergence of a major dominant software vendor has lent stability and more predictability to systems. The counterweight is that the pace of technology advance may be slower, but it is still substantial.
- *Client-server, Internet, extranet, and intranet development.* Where these projects were once rare, they are now becoming much more common and the focus of most of the new development. Such work results in large, complex projects to reduce risk.
- *Changes in staffing mixes.* The dominant programming language among staff was once COBOL. Although COBOL is still in substantial use, other technologies have become popular, such as C++, Visual Basic, Java, and fourth-generation languages. Students and people entering the programming field are far more likely to know these newer languages than they are to know COBOL. This influences what systems projects are undertaken and how they are carried out. With a greater variety of skills needed, project teams have to be larger. However, with an increased number of projects, the staff must be spread across multiple projects.

TRENDS IN PROJECT MANAGEMENT

These trends further influence the nature of systems and technology projects. Overall, the trends have the following impacts:

- Projects involve more business aspects due to reengineering, process improvement, and e-business.
- Projects are more complex in general in terms of technology.
- Projects are more integrated with each other and with current work.
- Staff members have to be assigned to multiple projects due to limited budgets and resources. In addition, there is competition for staff time between projects and normal work such as network support, maintenance, enhancements, and operations support.

More tools are used for project management than were available 15 years ago. Some highlights are the following:

- Project management software is now network based to support sharing of project information among project teams and with others outside of the project.
- Groupware allows for project teams to share information on issues, project status, action items, and other project information.
- Electronic mail, voice mail, and videoconferencing support the dissemination and sharing of information more rapidly.
- Databases support the storage and organization of information related to projects.

Note that these all enable sharing of project information and collaborative effort.

COMMON SYSTEMS PROJECT MANAGEMENT MYTHS

The preceding trends help to explode some of the myths that have arisen around project management over the years.

- **Myth 1: To be successful, projects require dedicated resources.**
  This was true 20 years ago. It is now no longer the case. Systems projects require a variety of different skills from different people over the life of the project. For most companies, having any substantial number of full-time staff dedicated full time to the length of the project is not possible.

- **Myth 2: Project leaders must be allowed a great deal of autonomy to get work done.**
  This myth is based on the idea that projects have a stand-alone nature and that project leaders are fully responsible for the project. Today, this is not realistic. Projects are interdependent; resources are shared. Project leaders may even be shared and split among multiple projects.

- **Myth 3: Success in a project is to achieve the implementation of the system or technology.**
  This is the traditional narrow definition of success. It is why many systems people think that a project was a success, while the business managers view it as a failure. It is a matter of perspective. Many systems people think that the job ends when the software runs; however, business managers think that the project ends when the business processes are improved with the implementation of the systems. A fundamental point in this book is although much of the costs of a project occur in the systems aspect, the benefits largely are realized in the changes and improvements to the business. Thus, if a project

stops with the completion of the system, there may be no project benefits. If the business processes and organization do not change when the new system is implemented then there are likely to be few benefits, making the project a waste of time.

- **Myth 4: Projects depend on stable, defined requirements and technology.**
  The traditional system life cycle emphasized getting the requirements right. If this were done, then the design and development are more likely to be successful because they met the requirements. This is nice in theory and terribly wrong in practice. The business managers may not know what is needed until they see something. As they see and experience the systems and technology, their requirements naturally evolve and change. It is part of life. It is also true that you must plan for technology improvement and change in any long-running project.

- **Myth 5: You can measure a project in terms of percentage complete and budgeted versus actual results.**
  This is the traditional and accepted approach that students are taught in project management classes. Unfortunately, in systems projects it is difficult, if not impossible, to clearly identify the percentage complete. In addition, much of the expense in a systems project is in fixed costs associated with networks, hardware, and software. The variable that mostly indicates problems is staff time. Thus, budget is a trailing indicator. You will see better measures for a project in terms of assessing outstanding issues and problems.

- **Myth 6: Each systems project is unique.**
  This is a seemingly innocent statement. However, if you accept it, then you tend to manage projects independently and fail to gain from lessons learned. Projects have unique attributes, but underneath the veneer of the new technology they have much in common.

- **Myth 7: You can throw resources and money at systems projects to good effect.**
  In some business projects this is true. It is seldom the case in systems management. People's skills and knowledge are not interchangeable. Throwing people into a systems project often just slows the project down due to coordination and bringing people up to speed.

- **Myth 8: Things will work out all right if you use the magic tool or software.**
  Any tool requires a substantial learning curve. A tool expert should be identified and on board. People not only must gain familiarity with the tool, but also expertise. Were this not enough, many tools are not complete. Staff members are forced to deal with multiple tools from different suppliers, making the job worse.

WHY IT PROJECTS FAIL

TRADITIONAL PROJECT PROCESS

From lessons learned, a major cause of failure occurs when an organization maintains the old project management process and methods to manage modern-day projects. Let's consider the problems with the traditional project management process. In general, the project management process can be divided into the following parts:

- **Ideas and proposals for projects.**
  In most organizations, ideas for projects are often expected to come from the business. The assumption here is that business units know their needs and so will formulate their requirements. These will then go to the systems organization for review and estimate of effort and schedule.

- **Project approval.**
  Several approaches are employed for approval. One is to use a steering committee that meets to approve or change priorities recommended by information systems. Although it may appear to be a slate of related projects, this is often a list of individual projects. Another approach is to negotiate and approve, defer, or reject requests individually. This is the "let's make a deal" approach.

- **Project kick-off and start.**
  Because the elapsed time between initial request and start of the project can be long, the project may get off to a very slow start. Generally an extensive planning effort is launched to develop a detailed project plan. Then this plan must be approved. Individuals must be signed up to be on the team. After considering the initial idea for the project, the idea is often found to be incomplete and more work must be done to define the scope of the project and end products. The project leader and team are often dedicated to the project. They have not been told to interface with other projects. That will, it is hoped, occur later. The project leaders are left to define their own tasks and milestones for each project, resulting in little or no uniformity across projects.

- **Project monitoring, reporting, and management.**
  Many people make a big deal out of the size of the project in terms of project monitoring and reporting. Larger projects get more attention; smaller projects get less. Project managers are often trained to focus on administrative matters such as tracking, reporting, and estimating for the project. Historically, several factors contributed to this. Project teams were larger. The software and other project management tools were primitive so that much of the project manager's time was consumed gathering status and using the software. Even today, there is usually no organized approach to deal with

problems. The project leader uses his or her own experience and knowledge to address the problems one at a time as they appear.

When it comes to review, managers consider the projects one at a time. They do the same with the problems and issues. More attention is given to projects that are over budget or behind schedule.

- **Project end or termination.**
The project progresses toward its milestones. At the end of the project, the business unit that wanted the project is asked to sign off. This can be a rushed affair. Neither the business unit nor the systems unit want the project to appear as a failure, which would reflect badly on them both. After sign-off, it is too seldom the case that management insists on determining the benefits of the system. If asked, the response is often in terms of soft, fuzzy benefits. Normally no effort is made to extract lessons learned from the project. The business unit goes back to its normal work. The system staff moves on to another project.

TWENTY-SIX REASONS THAT SYSTEMS PROJECTS FAIL

Many flaws are in the traditional approach that contribute to failure. Let's now build a list of 26 reasons why systems projects fail. These will be discussed in more detail in later chapters.

1. *Informative technology is reactive.* The information technology organization responds in a reactive mode to project requests. There is little or no proactive effort to search out good projects with major benefits.

2. *There is a lack of a project strategy.* Without an overall plan, many of the projects submitted and approved may provide only marginal benefits. Major projects involving the underlying technology and architecture infrastructure are ignored or viewed as impossible because no one business unit supports them.

3. *Project ideas are not well organized.* When you open up project idea development to any department, you are likely to attract those who are most interested in automation. These people tend to be younger and to manage smaller departments (due to less seniority). As a result, the project ideas sometimes do not stem from a clear business need, but rather from a desire to acquire a benefit that would be nice but may not be necessary.

4. *Projects goals are not understood or agreed on.* There is sometimes a lack of understanding and agreement about the goals of the project. Business, technical, and organizational objectives might overlap and conflict.

5. *Small projects are not allowed to be projects.* Because of the fear of administrative overhead and the desire for results, management allows small projects to exist without the controls and direction associated with project management. As

the scope of the project grows, the stage is set for collapse because no organized structure is in place. This occurs in business units that attempt to carry out projects informally.

6. *The business unit role is ill defined.* The role of the business unit is ignored or is assumed to be known. As a result, the business unit staff members struggle to find out what they are supposed to do. With few responsibilities, they are less committed to making the project successful. The potential tasks that they can address in terms of business process are not included in the project.

7. *Work is being done on the wrong projects.* The information systems unit is reacting to business units individually. Thus, many projects that involve critical business processes that span multiple organizations are left out because there is not one owner. Resources are consumed by intradepartment work.

8. *Projects are not closely linked to the business processes.* Projects are carried out with the attention on the technical side of the puzzle. The business process side is too often ignored. No tasks are related to reengineering. This can also be a problem with too narrow a scope.

9. *Benefits are not as they seem.* The benefits cited in the project are often estimated based on only the systems side without any business process change. Thus, often very few actual benefits are achieved.

10. *The scope of project is not defined.* The scope of the project is not well defined until substantial work has been done. Additional requirements that surface then enlarge the scope.

11. *The purpose of the project does not match the scope.* The goals of a project can be ambitious as reflected in the estimated benefits. Yet the scope of the project is too narrow, making it impossible to reach the goal.

12. *Projects are approved individually based on individual benefits.* This results from direct negotiation between the manager of systems and the business unit manager. This practice tends to encourage stand-alone projects, reducing the possibilities of economies of scale. This also makes the process of review political and biased toward specific business units that are on good terms with information systems.

13. *Projects that are approved by a committee are often approved based on fairness.* It is logical that the approved projects represent the desires of the committee members. Fairness says that each area receives some project support. This dilutes resources across many projects and turns them away from more important projects.

14. *Projects adopt tools without methods.* A tool supports a method. Without the method there is less likelihood of getting all of the value out of the tools.

15. *Projects are too rigid in either rejecting or accepting new technology during the project.* In some projects, there is an overwillingness to adopt a new tool—particularly if the project is in trouble. Adopting a tool then can just mask the underlying problems. In other projects, there is a rigidity to anything new or improved. In either case, the project is affected adversely.

16. *There is a lack of review of project ideas and plans.* No thorough and formal approach to reviewing a project idea or plan may exist. Often, the project is only reviewed by itself in a vacuum.

17. *The project team is weak.* The project team is weak or lacks technical knowledge and experience. No provision is in place for assigning senior people to the project.

18. *The project team was identified and gathered too early.* The project team members are identified and committed too early in the project. This leads to their under utilization. Morale can be depressed.

19. *Resources cannot be managed or shared across projects.* This is due to several factors. First, the resources are often dedicated to single projects. Second, when a high-priority task appears, the project team is stripped to handle the emergency—delaying the project. Third, no mechanism has been established to manage resources across multiple projects.

20. *Project resources are robbed to handle emergencies.* This occurs in non-systems projects as well. The problem with systems projects is that the only apparent downside of taking resources from a project is a delay in the project. Because the benefits are not clearly identified and the business unit is getting along without the systems or technology, there does not seem to be much harm. Don't kid yourself. There is. The people left on the project may work slower and may feel that they are working on a low-priority effort.

21. *The project manager spends too much time in administration.* The project manager is assumed to be carrying out the administrative tasks identified earlier. In many organizations this is assumed to be sufficient. The project manager may lose touch with the actual work on the project and fail to build rapport with the team.

22. *Issues are inconsistently handled.* Because issues are handled in an ad hoc manner based on the style of the project manager and his or her experience, analysis, results, and actions related to problems vary widely.

23. *Tracking multiple projects is difficult.* If each of five projects uses a different task plan with different task descriptions for the same work and different resource names, it is almost impossible to put the five schedules together for analysis.

24. *The mathematical critical path receives too much attention.* As you know, the critical path is the longest path in the project such that if any task on this path is delayed, the project is delayed. Who says that this path contains the tasks with greatest risk? It usually does not.

25. *Few or no lessons are learned.* With no formal or organized process to learn from the projects and a lack of incentives and management support, it is not surprising that few devote any time to gathering, organizing, or disseminating lessons learned.

26. *Lack of resource management between projects and regular work.* The problem here is that there is a lack of active management of staffing resources between normal work and project work.

GUIDELINES FOR SUCCESS

Let's turn from failure to success. Take the same phases of a project management process and consider a more modern method. This is the core of the approach in this book. All of the concepts that are discussed here will be expanded later. Here you will see the overall road map.

MODERN PROJECT MANAGEMENT PROCESS

- **Ideas and proposals for projects.**
  Although departments can still submit project ideas, information systems units work with departments to assess the current state of key business processes and their supporting systems and technology. The result is a *process plan* that leads to projects that involve process improvement as well as systems work. A complete *project concept* is developed that identifies business roles, purpose, scope, general costs, benefits, general schedule, and issues to be addressed in the project. The current processes are measured to assess the problems and to support the determination of benefits.

  The information technology function plans how to manage resources across multiple projects. The objective is to form project teams that are as small and flexible as possible. Analysis is also performed to identify and analyze the interdependencies between projects. Improvements to the architecture and technology base are identified and related to the business processes.

- **Project approval.**
  A *slate* of projects is reviewed and approved by management with an emphasis on interrelationships among projects. Project approval is tentative and subject to review at three-to-four month intervals.

- **Project kick-off and start.**
  Projects are initiated by developing detailed project plans based on standard templates. A *template* includes standardized high-level tasks, resources, resources assigned to tasks, and dependencies. This is more than a *work breakdown structure (WBS)*, which is a general list of tasks. It paves the way for increased standardization and the ability to analyze data from multiple projects. Team members are placed on teams in a just-in-time mode. The core team for any project is no more than three or four people. Other team members are on the projects part time for performing specific tasks. Lessons learned can be linked to tasks in the template. Issues can be linked to tasks in the project plan that is created from the template.

- **Project monitoring, reporting, and management.**
  Standardized electronic reporting is employed for all projects. A *project coordinator* analyzes the data from all projects. This person also reviews the active

issues across the projects and presents the results of the analysis to the project leaders as well as to management. The project coordinator facilitates the release of resources from projects and their redeployment to other projects. A project leader or IT manager can act as a coordinator.

Project leaders share their schedules with each other in a network environment. They work together in joint meetings chaired by the project coordinator to resolve conflicts for resources.

Project leaders do work on the projects. They also work on all active issues and support the gathering and dissemination of lessons learned. Project updating is done by team members through the network, releasing some of the project leader's time from administration.

Measurement of the project is based on open issues and their impacts on the projects. Percentage complete and budget status are still important but are considered secondarily.

- **Project end or termination.**
Although a project can end, it is usually just the end of a phase to support the business processes. As you know, making large-scale improvements in major systems involves creating and managing multiple projects.

WHY PROJECTS SUCCEED

Based on experience from many projects, there are other ingredients to project success.

- *Requirements.* Requirements in the modern setting come from the changes to the business processes. Business processes tend to be more stable than requirements created in the old manner. Requirements are often tied to business transactions. The requirements have less change and more stability. Moreover, any changes to requirements can be evaluated and justified based on an assessment of their effect and impact on business processes.
- *Team building.* By having team members more active in project management, they have a greater sense of commitment. The same is true for business units. Project leaders are encouraged to work together in a cooperative mode.
- *Issue oriented.* This process makes management more issue oriented rather than status oriented. Issues that cannot be resolved by individual or multiple project managers are considered by management.
- *Overall focus of projects.* Because the projects are based on major business processes that cross multiple departments, the projects will consume most of the available resources. This means that many of the smaller, enhancement type projects will fall by the wayside due to a lower priority.

- *Impact on information technology.* With this new project focus, information systems becomes a key supporter, if not owner, of processes that involve multiple departments.
- *Overall resource assignment.* Resources are actively and regularly allocated among projects and normal nonproject work.

ORGANIZATION OF THE BOOK

The organization of this book reflects the modern approach defined in this chapter. Part I focuses on the project management process. Here you will view the details of the process and get help and guidance in transitioning to the new process. How you can better manage multiple projects is considered here as well. Most books might not consider this factor because their attention is drawn to single projects.

Part II of the book deals with developing project plans, assembling the project team, and defining the new roles of the business units as well as the project leader. The development of the project concept prior to the project plan makes up an important chapter of this part.

Tracking, monitoring, and managing the projects are the thrust of Part III, along with an analysis of specific types of projects. Four of the major project areas are examined: software development; technology projects, operations, enhancement, and maintenance; software package acquisition and implementation; and technology projects. Part IV deal with issues in human resources, technology, business, vendors, and management.

With the exception of the last part, all chapters follow a parallel structure with an introduction, the approach, examples, e-business lessons learned, guidelines, and what to do next after reading the chapter. Part IV is organized around issues. The chapters have been written in a down-to-earth style so that you can put the materials to work for you and your organization immediately.

E-BUSINESS LESSONS LEARNED

E-business projects are more complex than standard IT projects for several reasons.

- The IT part of the project is more inclusive encompassing network, hardware, software packages, and software applications.
- Setting up or modifying business processes to support e-business is a critical success factor
- Organization change is often part of an e-business project.
- There is a need for e-business to be externally focused toward suppliers or customers.

In each chapter, lessons learned from both failed and successful e-business projects will be discussed.

WHAT TO DO NEXT

1. A logical first step is to assess the current project management process in your organization. Here is a list of questions to consider:
   - Is there a formal process for managing groups of projects?
   - Is each project managed differently based on the individual project manager?
   - Is there any effort to employ standardized project templates?
   - Is it possible to integrate several schedules and perform analysis?
   - Is there any effort to gather lessons learned from past and current projects?
   - How are project ideas generated? Are the same people always generating new project ideas?
   - How are projects approved?
   - How are benefits of projects determined?
   - Is there any effort to measure a project to ensure that the benefits are achieved?
   - Is there a standardized project reporting process across projects of all sizes?
   - How many of the current, active projects relate to key business processes?
   - How many business processes lack any projects?
   - Can people be moved between project teams?
   - What is the size of project teams?
   - What is the role of the business unit staff in projects? Is it consistent across multiple projects?
   - How are issues defined and addressed? Is there an effort to see which issues apply across multiple projects?
2. Consider some recent projects that were failures. Ask the following questions:
   - Was there an effort to learn from the failure?
   - Did the project management process change as a result of the failure?
   - At what point did people know that the project was a failure? How long was it before the failure was acknowledged?
   - How was the failure announced?
   - What was the reaction among team members of other projects?
   - What were the dependencies between the failed projects and others?
   - What major issues contributed to failure? How long had they remained unresolved?

3. Consider some recent projects that were successes. Ask the following questions:
   - Was the project only a technical success?
   - Was success announced at the conclusion of the project or after the new system had been in place and measured?
   - Were any lessons learned from the success?
   - What were the major factors that contributed to success?

SUMMARY

The road map of this book and approach has been laid out. It is a modern approach that relies on the network technology of today and the future. The thrust is on empowerment and effectiveness in the new downsized world of information systems and business. The approach is much more related to the business environment than to technical factors. In Figure 1.1 you can see that business issues relating to business processes can be translated into systems issues and the systems and architecture. Overall, you seek to mesh the objectives and strategies of systems with those of the business through the link between systems and business processes.

Chapter 2

Developing Your Project Management Process and Strategy

INTRODUCTION

The purpose of this chapter is to provide you with guidelines to

- Evaluate your current project management process.
- Develop a project strategy.
- Determine improvements to the process.
- Transition the current project management process to an improved one.

What is a *project strategy*? It is your overall approach for setting up and managing projects. What you decide to do here will be carried out in your new project management process. All of your projects will then be managed within the process and follow the strategy.

To better understand what a strategy is, let's consider some alternatives.

- **Alternative strategy 1: No strategy at all.**
  Here project ideas are generated and projects are started on an individual negotiated basis. Each project is managed and measured on its own. Interfaces are very difficult because no collaborative culture exists. It is everyone for oneself. You are unable to combine the data in different projects or gather lessons learned across projects.

  Without a strategy, the entire process of managing the projects can become disorganized. Results will be at best uneven. Because there is no overall vision or view, there is more likelihood that major projects will be missed or left out. Also, critical infrastructure projects to establish or improve networks will be more difficult to approve because they lack a specific business department client or owner.

- **Alternative strategy 2: Carefully select which projects are funded, but then manage each separately after they start.**
  Many organizations have drifted into this approach. Management has become aware of the importance of projects, so the project receives a great deal of management attention at the front end. Unfortunately, this attention is not carried through during the project. There is often an underlying assumption that if you start the projects out correctly, then the project managers and teams will see them through. This is not the case with systems and technology projects that are highly interdependent. Issues that are unresolved in one project can quickly impact other projects. How will upper management become aware of the impacts of such an issue? If no mechanism is in place to deal with issues, they can mature into crises.

- **Alternative strategy 3: Proactively determine which projects are funded based on tangible business benefits and then manage the projects collectively.**
  This strategy assumes that the whole of the projects is greater than the sum of the individual projects. This strategy results in a collaborative management approach that involves the management of issues, resources, and schedules across multiple projects. As you can imagine, this consumes more management time and effort in dealing with issues. Results and project success in a number of firms indicate that this is the approach to consider for your organization. Benefits can be seen tangibly in the business process improvement. This applies in e-business where the implementation is very large in scope so that e-business implementation consists of a number of projects.

You probably have most often observed the first two alternatives. However, from the previous chapter you learned that substantial benefits can be derived from the third alternative. Thus, restating part of the purpose of the chapter, you seek to find ways to migrate from alternatives 1 and 2 to alternative 3. Why haven't more firms adopted more aspects of the collaborative strategy? Experience reveals many reasons. First, the organizational culture may reinforce the old systems mentality of separate projects (dating back to stand-alone batch and limited on-line systems). Culture is hard to change. Second, if you work on maintenance and fail to take on major projects, then you can avoid developing a strategy. Third, there may be a lack of tools and support for shared work.

APPROACH

A project strategy leads to defining the project management process. The *project management process* is how you manage all of the systems projects from their conception through completion or termination, as was informally discussed in

Chapter 1. If you assume that the third strategy applies, then the project management process includes the following goals:

- Ensure a proactive approach in identifying potential projects that link to key business processes.
- Encourage wide participation and commitment by business units in projects.
- Obtain upper-management commitment by setting the slate of projects and dealing with major issues that cross projects.
- Enforce uniformity in project plan structure through templates, common databases for issues, action items, lessons learned, and standardized reporting.
- Support information sharing among projects and resolving issues among project leaders.
- Provide analysis of multiple projects and dynamic resource management across projects so as to spread resources effectively across projects.

Now let's make this discussion more formal by putting it into a series of steps. You will be considering each step in the context of the more proactive project strategy. For each step the end products and key questions you must address will be defined.

STEP 1: ESTABLISH BASIC COMPONENTS OF THE PROJECT MANAGEMENT PROCESS

The end products are as follows:

- Approaches for managing multiple projects, project reporting, issue management, and lessons learned are defined.
- The project management process (defining what is to be done in each of the nine steps) is documented, presented, and approved.
- A steering committee is appointed.
- A project coordinator is identified.

It is important to define the role of the project coordinator(s). The *project coordinator(s)* is responsible for the following activities:

- Assisting business departments in developing process plans
- Supporting the development of project concepts
- Providing support for training in project management methods and tools
- Analyzing project concepts and recommending a project slate to the management steering committee
- Assessing project issues and status and presenting analysis results to management

- Coordinating the identification, analysis, and resolution of issues
- Supporting the lessons learned database

A project coordinator does not have to be a full-time role. It can be filled by a line manager or on a rotating basis by project managers.

Whether the project manager is one person or several depends on the projects, organization, and range of business and system activities. Note that these responsibilities are similar, but expanded from the role of the traditional concept of a *project office*. In the past, the project office typically concentrated on the tracking activities of systems projects and then provided analysis results and reports to management. A project office performs the following functions:

- Provides a centralized review of all projects
- Reports on projects to management
- Sets standards and procedures for project management
- Supports project management software

For some companies, the project office has provided a useful way to gain control over projects and provide uniform reporting. A survey indicated that 15% of firms have established project offices. However, by its very nature the project office is pure overhead. It can take on a life of its own and impose major burdens on projects. The project office can also separate project management from project administration.

What are the attributes of a project coordinator? He or she should know project management and see the opportunity to improve the current process. The project manager should also be able to perform a number of tasks in parallel. The candidate does not need to be an expert with project management software and does not need to be technically oriented, because this role will entail working closely with business processes, business units, and business department staff.

Let's consider some defining questions. What will be the general scope of projects? You can arrange alternatives for project scope into levels. Your goal is to have as many projects at higher levels as possible to support the business.

- *Level 1.* The project is restricted to the system itself. This is very narrow and traditional. There is no guarantee that either architecture or business process improvements will be carried out.
- *Level 2.* The project includes infrastructure or architecture along with the system. This is a more modern view and reflective of network-based systems such as client-server or intranet projects as well as data warehousing projects that often involve database management and hardware upgrades.
- *Level 3.* The project includes the system, architecture, and business process improvements. This is the comprehensive scope that is most desirable. Although this scope is larger, you can divide the project into subprojects.

What is the role of the business units in projects? This is a crucial question. One part of the answer is the involvement of high-level business managers in the management steering committee. Another part consists of the following activities, which are performed by business units:

- Developing process plans—blueprints for the long-term future of the business process
- Participating in individual projects
- Supporting process change linked to new systems projects
- Insisting on measurement of projects—during and after work

How should the *management steering committee* be constituted and operated? The steering committee consists of high-level business managers whose duties include the following:

- Reviewing and approving the slate of projects
- Assisting and resolving project issues that cross multiple projects from a business perspective
- Periodically meeting to review and reset project priorities based on current needs and the state of the projects

The committee can meet from two to four times per year based on issues and the budget cycle. Note that if it meets more often, managers will tend to send lower-level replacements. The agenda will start drifting into considering project status. Then the steering committee ceases to be a steering committee and becomes a project oversight group—another layer of red tape.

When thinking about the management steering committee, you must consider the budget cycle. Let's suppose that your organization operates on a calendar year. This means that the first cut of the budget will be completed in October or earlier. The project slate must be in place prior to this date. If you back up from October, you might select August or early September as the time for your first meeting. Assume that there are to be four meetings a year equally spaced through the year. The second meeting might be in December, during which time results and issues are to be presented. A third meeting might be in March. This is an important meeting during which project priorities might be realigned. The fourth meeting might be in June and resemble the December meeting. If you think that this is too much detail, you are wrong. The more you can plan and make known to management, the more likely you are to attract the right people to the meetings.

Who are potential candidates for the committee? Assuming that you have identified several major business processes that merit projects, then the business unit managers most involved in these are logical candidates. When selecting between line managers and staff managers, choose the line managers. They command the attention of top management and their support will be critical.

Why do you have to have a formal method for addressing issues? Why not address each issue as it arises? This is the seat of the pants of the reactive approach. It will yield inconsistent actions and sometimes raise even more issues. Proactive issue management aims to track and analyze all issues that have remained unresolved for some time. Issues management also attempts to determine the impact of issues and their resolution across multiple projects. Because risk in tasks and the plan are reflected in the issues, issues management covers traditional risk management in projects.

STEP 2: DEFINE CRITICAL BUSINESS PROCESSES AND ACTIVITIES AS WELL AS THE SYSTEM ARCHITECTURE AND PROCESS PLANS

In order to be proactive, you must understand what areas are crucial to both the business and the systems in terms of infrastructure and architecture. This is the first step in detailed analysis. The step can help management and the business units focus on what is critical overall. Isn't this done naturally? Not really. Many general managers are uncomfortable with systems. The business processes cross the boundaries between business units; business unit managers may be reluctant to address issues outside of their control.

The end products of this step are as follows:

- Definition of system architecture
- Identification of critical business processes
- Process plans for each critical process

Critical questions here are as follows:

- What is the current systems architecture? What improvements are needed based on current and known projected systems requirements?
- How will critical business processes be determined? You will not be able to generate projects for all processes. In addition, resources will still be required for maintenance and operations support as well as architecture improvement.
- What constitutes a process plan? How will process plans be developed? These questions will be addressed in Chapter 4.

STEP 3: REVIEW ALL CURRENT PROJECTS AND ASSOCIATE THEM WITH THE PROCESS PLANS AND ARCHITECTURE

This can be an eye-opening experience. You may find that many of the projects do not fit with the business processes and architecture. They may instead be directed toward enhancements and maintenance or minor systems development. The importance of this step lies in showing management and the business units that the way things are currently being prioritized needs to be revisited. This can help set the stage for the new project management process.

The end products are as follows:

- Assessment of current projects along with necessary changes and recommendations in relation to the process plans and architecture
- Recommendations on which projects can be terminated due to limited benefits

It is in this step that the shortcomings of the current projects are identified. From lessons learned, several problems can occur:

- The scope of the current project is too narrow and does not include business processes.
- The business role in the project is not defined.
- There are few, if any, tangible benefits to the project as it stands.
- The interdependencies between projects have not been completely specified.
- When you add up all of the projects, there are insufficient resources and substantial unplanned work.

As a result, you will likely recommend that some projects be killed, others restructured, and still more combined. Killing a project has several benefits. First, resources are released to work on other projects. Second, management shows that it is serious about insisting on project results. Third, the organization has an opportunity to learn lessons from the experience.

STEP 4: DEVELOP PROJECT CONCEPTS FOR NEW PROJECT IDEAS AND CURRENT PROJECTS

This positive step counteracts the negative previous step. Do not present the results of step 3 to management without step 4, balancing the negative with the positive. The project concept provides sufficient information to build management support. The project concept nails down the purpose, scope, general schedule, benefits, and costs. Why have a project concept and not a detailed project plan? Because the project plan assumes a specific scope and sets too much of the project in concrete. The concept provides a vehicle to negotiate with management on the scope and relationship among projects, for example.

Why have a project concept for an existing project? The project concepts for existing projects begin to reformulate the projects themselves and provide greater focus from a business perspective. Many surprises may be revealed here. Some encountered in the past include the following:

- The scope of the project had changed and no one had informed management.
- There were hidden or assumed dependencies that now surfaced, which resulted in schedule delays.
- Business unit support evaporated for several projects when they were faced with identifying specific benefits.

The end products are (1) project concepts for new project ideas and (2) project concepts for current projects.

STEP 5: ANALYZE AND RELATE ALL PROJECT CONCEPTS

As was stated earlier, projects do not stand alone. The whole is greater than the sum of its parts. In this step you will determine how the projects relate to each other, what gaps have to be filled, and the additional benefits that will accrue if multiple projects are implemented. In this step you will flesh out the project concepts. This means that you are more likely to have a complete view of the benefits of the projects in total. This can prove useful later when you present management with the project slate.

Some examples of results of the analysis are as follows:

- A business unit was going to be using two client-server applications. Hence, a common, standard user interface was required. This was unanticipated before the analysis.
- The wide area network had to be expanded to accommodate additional users.
- Several supporting projects of lesser priority now became higher priority due to the dependence of high-priority projects on them.

The end products are as follows:

- Analysis and determination of relationships and dependencies between project concepts
- Adjustment of project concepts to cover gaps identified in analysis
- Analysis of resources across project concepts
- Recommended prioritization of project concepts

STEP 6: DETERMINE WHICH PROJECTS WILL BE APPROVED FOR IMPLEMENTATION

With the analysis of step 5, you are ready to present management with groups of projects. You must avoid wherever possible reverting back to single projects. When you present and market a group of projects, you can enlist the support of multiple high-level managers of business units. These managers will not feel that they are part of a zero-sum game, in which one party wins while another must lose. The accumulated benefits of the group of projects can be presented. The end products of the step are as follows:

- Approved slate of projects
- Project leaders identified for all projects

STEP 7: DEVELOP DETAILED PROJECT PLANS FOR APPROVED PROJECTS AND BEGIN WORK

In traditional systems project management, step 7 might have been step 1. It is not the case here. With management approval in step 6, you are ready to develop the detailed project plans and identify project leaders and key team members. The thrust is to give attention to interfaces and other areas of risk from the beginning. Different rules for building the project plan will be defined. The end products are as follows:

- Detailed project plans for all projects
- Refined budget and schedules for all projects
- Key team members initially required for each project
- Management review and approval of all plans

STEP 8: MANAGE, MONITOR, AND REPORT ON THE PROJECT SLATE

In traditional project management, this step normally involves reporting on status and the percentage complete. That is only part of the puzzle and not very revealing at that. Here you will concentrate on three major activities: managing issues, allocating resources across projects, and evaluating actual work results from the projects. This is much more proactive than status reporting.

A wide range of end products are possible. Issues are resolved. Project plans can be changed. Priorities can be reset. The new project management process gives a much more active role to management and the project coordinator in addressing issues and actively overseeing the projects.

Part IV of this book contains six chapters on issues. Why six chapters? Because this is a book on how to do technology project management. Issues of all types are addressed in this part of the book. These issues are very specific and tend to recur in various forms in different projects. You must know how to recognize the symptoms of these issues, how to analyze and group the issues, how to market for a decision, and how to implement actions that support the decisions.

STEP 9: DEVELOP AND IMPLEMENT CRITERIA FOR PROJECT ENDING AND TERMINATION

This is an unpleasant subject that people avoid. It is essential to have termination in mind for almost all projects. You cannot afford to have a failing project drift along. By taking a more proactive approach, you can redirect projects, change their scope, and salvage failing projects. Because projects are more interdependent, the importance of salvage cannot be underestimated. If you manage the issues actively in the previous step, you can start to see the impact of unresolved issues on the project earlier. This will give you time for action.

Critical to killing or abandoning projects is an organized approach for salvaging lessons learned and even end products. How do you do this if you have never had experience? Take a project with which you are familiar and assume that it were to be killed. What can you salvage that might be useful? What would the business unit do if the project were to be killed? What would its fallback position be?

TRANSITION TO THE NEW PROJECT MANAGEMENT PROCESS

With the steps defined, you might wonder how a systems organization can migrate to the project strategy and process. Experience shows that management must see benefits from the potential of a new approach before it will be considered seriously. Here are some suggestions:

- Evaluate your current projects and associate them with major business processes.
- Examine each project to determine if it could be changed, reduced, or eliminated.
- Determine what resources can be freed up if the previous step were taken.
- Develop a model or scenario of a list of projects that you think might yield the most benefits.
- Identify the general benefits of the projects in terms of example business transactions.
- Define comparison tables as discussed later.
- Develop a project plan for making the transition to the new project management process.
- Present the results informally to management to trigger interest.

The differences between the old and new process are revealed through project slate change and the benefits of the new projects on the slate. Here are descriptions of some comparison tables that have proved effective in the past.

- *Potential new projects versus existing projects.* The table entry consists of comments on how the new project relates to the existing project and reveals how different the new projects are from the existing ones.
- *Potential new projects versus business processes.* This table shows how the new projects support the business processes. Each entry explains how the business is supported by the project.
- *Existing projects versus business processes.* Similar to the preceding table, this table shows the limitations of the current project for the specific process. Comparing this and the previous table can be enlightening.
- *Potential new projects versus business units.* This table explains how each new project will benefit the major business units.
- *Potential new projects versus benefits.* This table explains how each new project will benefit the major business units.
- *Potential new projects versus benefits.* The columns are types of tangible benefits. Each entry explains how the project will yield the benefit.
- *Potential new projects versus potential new projects.* This table can be employed to display and explain the interdependence among projects. Each table entry explains how the column project supports the row project. This information is valuable for gaining support for architecture and infrastructure projects.
- *Existing and potential new projects versus issues.* Here you identify what issues the projects have in common. The table entry is a short description of how an issue applies to a project.

MARKETING THE NEW PROJECT MANAGEMENT PROCESS

Why market the process? Can't a logical new method sell itself? Hardly. A number of barriers need to be overcome, including the following:

- *Internal resistance from the systems staff.* Programmers who are comfortable with maintenance and operations support may not appreciate or want to participate in larger-scale development or software acquisition projects.
- *Internal resistance from systems managers.* Traditional systems managers may like the current process. They will point out that it has worked for some time. Experience shows that they will resist including the business process in their tasks. They also feel uncomfortable with a wider role for the business units. In addition, they may feel that they have lost power in setting priorities.
- *Lack of interest among business units.* Although some business units may "step up to the plate," others will not want to assume any other duties. They

may not see anything in it for them by addressing a cross-department business process.

* *Lack of support from management.* Managers may be supportive—up to a point. When issues have to be addressed, their interest may wane.

The attack on the new approach can take many forms. A common tactic is to argue that changing the current projects will be too disruptive. The new process can be phased in later—like next year or beyond. How do you respond to this argument? Point out what potential changes you have identified with the current projects and what the benefits are. Also, point out what will be wasted if management decides to wait. Resistance is often not on the surface. The suggestion is for you to assume that factors such as these exist and that they are addressed.

To help you with marketing, consider the benefits of the new process to each audience. Here are some benefits that have proved useful in the past:

* Business unit managers
  —Work on their critical processes will be more intense and yield greater benefits.
  —They will have a greater role in setting priorities.
  —They can impact the scope of the projects.
* Upper-level managers
  —Reinforce the importance of projects to the organization and business.
  —Indicate that their involvement will be very limited.
  —They will be able to change priorities on a more controlled basis.
* Systems management and staff
  —They will be working on critical projects.
  —Business unit involvement will be more carefully defined.

How do you go about doing sales and marketing? The sequence used to approach people is important. If you do it wrong, you can create enemies. Here are some specific suggestions:

* Start with business unit managers in line organizations. They have the most to gain directly from the new process. They can help with their own staffs and with management.
* Approach the systems managers and point out the benefits of the new process to them.
* Spend time working with the systems staff to build support.

In making presentations, be sensitive to the audience and present the project slate to appeal to their self-interests. Avoid using technical terms. Clearly indicate the business reasons for dependencies between application-oriented projects and technology/infrastructure-related projects.

What can you do to get people to act now, rather than waiting? What is the price of waiting?

- Project resources will be wasted in the projects that you have tagged for cancellation and change. What can be done with these resources in the interim if action were to be taken now?
- Relationships between business units and information systems will be affected if known gaps, interface problems, and other issues are not addressed.

A key here is that once people see that there is a better way to manage the projects, then there will pressure to move to the new approach.

EXAMPLE: BEAUMONT INSURANCE

Beaumont Insurance is a standard homeowner and automobile insurance carrier. This example will be discussed in several chapters. Beaumont's key processes are insurance application processing, policy servicing and tracking, payment processing, accounting, and claims. The previous management of Beaumont had decided to establish a major project for application processing and tracking using CASE (computer-aided software engineering) tools. The systems part of the project team started with 5 people and then grew to more than 20. The business units involved committed more than 10 people full time to the project. The project had gone on for 15 months, yet the company had only a prototype to show for the more than $3 million in expenses. The impact of the drain of resources for this new project was widespread and severe. First, almost all other nonmaintenance work was halted. No enhancements were made to other systems. The architecture, which consisted of terminals and some PCs, was frozen to release funds for the project. Business unit managers constantly complained that they could not be competitive because they could not offer new insurance products. The investment in the system provoked a crisis in management. New management was brought in.

The first step was to assess the current project and other systems work. The hardware and network were out of date and too expensive. The large insurance application processing and tracking project was out of control. Efficiency in the business units was very low due to the use of the legacy systems and old procedures without the new system. It was clear that major change was required. All of the business processes identified earlier required major work along with system replacement. No suitable software packages were available that fit Beaumont's situation.

The decision was made to concentrate on those processes that contributed directly to revenue or that incurred large costs. This meant priority went to application processing, policy servicing and tracking, and payment processing. These were the bread and butter of the firm. Accounting was dropped from consideration due to limited resources and the fact that benefits were much less. The claims area was a possibility, but it was not a major problem. A plan was developed for each of these processes. This plan included procedures, business policies, systems, and the infrastructure. The three plans were then applied to the developed project concepts. The current large systems project failed the test when mapped against the project concept. The project might have resulted in a new system that might have been easier to maintain but would not have had added new features. Moreover, it did not employ client-server technology. The assessment was that there might have been no business benefits from continuing.

Given the decision to kill the project, salvage began immediately. Much analysis had been performed to define business rules or the detailed procedures of how the current business processes worked. This had not been done formally before and was useful for the future. What was to be the future? Much of the design had to be dropped. Extensive experimentation on various project concepts resulted in the following projects:

- *Project 1.* Establish a wide area network to replace the terminal network. The benefits were improved response time in accessing the legacy system as well as providing local functions that were now done manually.
- *Project 2.* Reengineer and improve the business processes with the current system as well as in preparation for the new system.
- *Project 3.* Create a client-server system in which the client side of the system was modern whereas the server side was really a front end into the legacy system. This would give the business staff increased productivity while providing interim benefits until the new client-server system was implemented.
- *Project 4.* Develop a new client-server application that employed the business rules from the failed effort, the client software from the previous project, and database technology along with some object-oriented rules for development.
- *Project 5.* Modernize the host hardware to support client-server and database management technology.

The management steering committee was established, and its members were very unhappy in their first meeting. They were told the extent of failure and what had to be done. Several business unit managers committed to the new projects immediately. They had no real choice. The managers of the units that were to be sidestepped were unanimous in their opposition. This was overruled and work began on all but the fourth project at the start. The infrastructure and client-server front end system would have to be in place before serious work could be performed on the new system.

The project coordinator role was dedicated to analyzing the business processes and doing the overall project management. Pairs of managers from systems and the business units developed project concepts and detailed plans for each of the projects. The project coordinator put the projects together and performed analysis.

This led to more dynamic resource allocation. Initially, resource allocation was reviewed every two weeks. When there was more progress, it was stretched first to monthly and then once every two months.

From the start the focus was on issues. The previous management had tolerated the project plan even though it was only in a prototype stage, but was listed as 70% complete (incredible, but true). The project coordinator worked with staff from the business and systems units to create a list of more than 350 issues. A number of these had nothing to do with systems but involved business procedures and policies. For example, one policy determined when to send out bills to customers. Another established when to terminate coverage for nonpayment of insurance premiums. Schedules were managed around the issues to ensure that the issues were addressed.

Overall, the project management process was successful. However, ad hoc requests from marketing and other areas often had to be deferred to focus on key requirements. For each request, the manager was required to provide information on direct benefits related to cost savings or revenue generation. This philosophy reduced the backlog of work requests by over 80%. More details of the Beaumont Insurance example will be given in later chapters.

E-BUSINESS LESSONS LEARNED

Many firms have rushed into e-business—often with disappointing results. Why does this happen? Among the reasons are that the organization had no project management approach and process in place that could address e-business implementation. Having turned around a number of efforts, we have noticed that the following problems are typical.

- The company treats e-business implementation as a narrow techology project. As such the emphasis is on installing e-commerce software, enhancing their network, and adding other software such as firewalls. They ignore changing marketing, purchasing, accounting, ordering, customer service and other areas.
- The project starts as a single large project. The project leader is often an IT systems analyst. He or she is well intentioned but treats the project as a standard IT project.
- Issues begin to arise in both the business and technical areas. There is pressure to implement e-business fast. Hence, there is a rush to deal with issues as soon as they come up—a big mistake. Decisions and actions on later issues undo those for earlier ones. The project loses focus.

Turning from problems to success, some critical success factors in e-business implementation are as follows:

- Recognizing that e-business implementation is not a one shot deal. It is a continuous program since once you set it up there is a drive to expand e-business and keep it competitive.
- Having several project leaders who divide up accountabilities and can address both the business and technical sides of the project.
- Setting the scope of the implementation to include process, procedure, policy, and organization change as well as software and infrastructure implementation.
- Dividing the implementation project into a number of parts. The division is based on responsibility with a focus on ensuring that task areas with issues and risk are given high exposure in the plan.
- Actively managing issues during the project.
- Gathering lessons learned and experience during the project.

GUIDELINES

- In completing the analysis steps outlined in this chapter, be careful to limit your visibility. The more visible the potential change in project priorities is, the more danger there is of an attack and confusion.
- If you think about it, the major differences between large and small projects lie in scale, scope, and purpose. There can be more risk in some small projects than in larger ones. Work to have all projects included in the analysis and prioritization. If small projects are allowed to be treated separately, then the resources available for allocation will be reduced. You become locked in.
- It is easy to set aside resources for maintenance and operations support. Often, the most trained and experienced staff members are required for these tasks. These activities can then escape being prioritized. With so little of the time of the key people available for new projects, there may be little point in having the projects go forward.
- Even with limited resources, still make a proactive effort to control resource allocation and limit work on current systems and the architecture. Otherwise, each year will just bring more maintenance requests.
- In considering a transition to the new process, you might want to use a department as a pilot case for a project. This will demonstrate the project concept, issue management, and analysis of projects that are parts of the new process.
- In order to test out the resource allocation across projects, take the projects that were approved in the previous year. Go through the analysis suggested here and identify what different priorities could have been set.

- There is resistance to change that we explore in the last chapter. Here be aware that a major source of resisting change is the desire to maintain autonomy and control. The paradox is that the project management process here gives structure at the highest level and flexibility in the detail and in dealing with issues—areas where it is needed.

WHAT TO DO NEXT

In Chapter 1, you answered questions about the current systems project management process. The following actions flow from the answers to these questions.

1. Identify the key decision makers and supporters you require to implement a new process:
   a. Upper management
   b. Staff to general management
   c. Business unit management
   d. Business unit employees
   e. Systems managers
   f. Systems staff
2. Following up on the discussion in this chapter, estimate what types of resistance you will encounter.
3. Identify several potential candidates for the role of project coordinator. Write down what you consider to be the duties of the project coordinator. This will depend on the number and extent of projects. Assuming that it was not a full-time role, what other duties can this person perform?
4. Identify potential members of the management steering committee based on the earlier discussion. Estimate what their schedule of meetings might be. Write down a list of things that they might be required to do. This includes making decisions on issues that only management can solve as well as redirecting resources.

SUMMARY

If the underlying method for managing systems projects is not effective, then being successful with a few projects may not be enough. The cost of a poor process is high—resources that are locked into projects, a higher likelihood of failure due to issues not being managed, and projects that do not deliver tangible benefits because they may lack links to business processes and support of business units.

A proactive systems project management approach will attempt to identify the most beneficial project ideas in advance of planning and setting the project slate. Once identified, there must be a step in planning that allows management to trade off alternative goals and scope. This is the project concept. Having upper management involved in setting the project slate and reviewing it periodically not only increases involvement and commitment, but also provides for a channel to handle ad hoc requests for new projects.

Chapter 3

Managing Multiple Projects and the Project Slate

INTRODUCTION

The new project management process centers on managing a portfolio of projects as opposed to single projects. How to analyze and manage multiple projects in more detail is a critical part of the process and the focus of this chapter. The remaining chapters will develop methods that fit within the framework.

This chapter explores how projects are interdependent. Then guidelines are provided for handling analysis, resource allocation and conflicts, issue resolution, and disseminating lessons learned. The scope will cover all of your projects. Do not omit any projects because they are small, large, or have some special characteristic.

What if you try to manage the projects separately? There will be less sharing of resources. Project managers will tend to work with one another only when an issue becomes major or a crisis. Management will have to be more involved because more issues will bubble up. There will be greater incompatibility between software because there is less coordination among members of teams doing similar development.

What are some of the benefits to managing projects as a whole rather than as individual units?

- Information can be shared among projects. This includes lessons learned and issues.
- Project managers can solve issues among themselves because a vehicle for supporting issue resolution is available through the project coordinator. Remember that the coordinator is just a role that can be filled by one of the IT managers.

- Resource shortages and overages can be adjusted by analyzing resource needs across all projects.
- The focus of attention can be placed on interfaces where the risk often lies in systems projects.
- The same methods and tools can be employed across multiple projects faster.

However, there will be resistance as well. Here are some examples by source:

- *Project managers.* Their independence is impacted in several ways. First, they must adhere to a standardized project template with the same resource list and high-level tasks. Second, they must share information with other project leaders. Third, they must contribute and participate in resolving issues and using lessons learned. Fourth, and perhaps more important, they must be prepared to give up resources.
- *Business units.* In the past, a business unit would be assigned a project after management approval. In many cases, the business unit managers felt that they could make changes to the scope, purpose, and structure of the project at will. After all, they controlled it. In an open setting, this control goes away because the project information is more widely known and being tracked. Changes in purpose and scope are subject to more intensive review.
- *Upper management.* In the past, high-level managers could reset priorities more easily because the project information was not as visible. When the portfolio is managed as a whole and information is visible, then the high-level manager must approach change with more caution.
- *Information systems staff.* Some people working on projects like to work alone on their specific tasks without having to spend time in coordination and planning. This is a luxury that systems departments can little afford. A programmer's work is more complex because he or she must consider the impact across multiple projects.
- *Vendors and consultants.* Vendors hired to work on a project often find it to be in their best interests to keep the project going and even expand it to increase the work. In the process identified here, the change in scope is reflected in the project tasks and schedule. This tips off management that there was an issue here.

How do you disarm resistance? Your ignore it at your peril. Involve project leaders in the definition of common issues and the template. A template consists of standard high-level tasks, resources, and dependencies. Then you can establish a process for sharing information. Sit back and see what happens. Will the managers cooperate on their own? For the business units and upper management, a suggestion is to indicate the impacts of past project changes in purpose and scope on multiple projects. Also, point out lost opportunities in past projects for cooperation that resulted in extra work. An approach for the information systems staff and vendors is to indicate how the nature of systems projects has changed and how interdependent projects are. Point to the challenge of trying to spread the results of work across multiple projects.

APPROACH

CATEGORIES OF INTERDEPENDENCE

How are projects interdependent? In a standard project management context, the answer is simple. Interdependence is determined by the dependence of specific tasks and milestones between several projects. In systems and technology projects it is more complex. Five categories of dependence are identified next. Two projects can be interdependent and fit in multiple categories. Which categories a project falls into will help determine how projects can be grouped.

Category 1: Technology Base

Systems projects here share the same technology base. Some examples are as follows:

- *Sharing the same network.* As an example, suppose that you were implementing a client-server system, upgrading electronic mail, and deploying groupware. These three projects share the same network. By themselves the workload in implementation and testing along with the load on the network is reasonable. Taken together, it is likely that the network will have to be upgraded. After adding this project to the list, you now must prioritize the network staff time among support of the projects.
- *Sharing databases.* Several projects involving an application can include a database. The basic application may be enhanced. There could be a data warehousing project that accesses the database. A client-server or intranet project may also require the data. If you attempt to do these in parallel without tight coordination, then you risk having to do substantial rework when the database is changed.
- *Using a common GPS and GIS system.* In transportation and other areas where geographic information is used by multiple applications, it is important to design the applications as a whole so as to avoid redundant and inconsistent data.

How will you address these issues in managing multiple projects? You want to make sure that a technology project is created and that all interfaces are covered across the projects. You might want to appoint a superproject leader for the group of projects.

Category 2: Resources

The most common shared resource is people. In client-server applications, it is logical to use the same programmers to develop all of the client software across multiple applications. It is the same with the design of databases. In intranet and Internet applications involving electronic commerce, you logically want one or several staff members to centerpost all electronic commerce applications—regardless of their nature. Network support staff members are also shared. Programmers who maintain legacy systems are often shared between these systems and new, replacement systems. This is because they are the only ones with knowledge. Quality assurance staff members have to be allocated across applications for testing as well. In an intensive development environment, the development and testing hardware, network, and software must also be allocated.

Resource analysis and allocation can be simplified if standard templates for the projects are in place. Then you can combine the projects and extract all tasks for the next two months that involve a specific scarce resource.

Category 3: Milestones

This is the traditional mode of dependency. Tasks in one project depend on milestones in another project. Note that many people have trouble with these dependencies because they do not control them. Some give them little attention and wait for the project leader of the dependent project to inform them of problems and slippage. This, of course, gives the project leader a chance to slip the schedule of the other project.

Lessons learned point to a better method of dealing with these dependencies. First, pull out all of the interfaces from both projects to form a separate project. When the project plans are updated, the new project plan with the dependent tasks and milestones can be updated. Second, have the project coordinator monitor these "interface projects." This gives special attention to interfaces and potential schedule problems.

Category 4: Business Units

Projects can impact the same business unit. Therefore, the projects can require the same business unit support staff. That is an allocation problem. To cope, insist that business unit resources be included in the resource list for the project and then be scheduled in the template. This will ensure that business unit resources are not bypassed or ignored. The business unit staff issues can be dealt with in the same manner as that of resources in category 2.

The next problem is that implementation of the systems created by the projects can cause havoc in the business unit. In one government agency, separate groups were upgrading the PCs, the software, and the network and loading an on-line application at the same time. It drove the staff crazy and caused the work in the department to grind to a halt.

Category 5: General Business Need

Your company can face a major competitive threat. It may be forced to implement a number of major projects at the same time. These systems then affect major parts of the business concurrently. The most effective approach is to establish a superproject that includes tasks from the business viewpoint. This will give attention to the impact on the business effects of the systems.

A common theme runs through the discussion of these categories. It is that the project information must be combined, filtered, and extracted to suit the specific category of dependency. Fortunately, with modern project management software this is easier to do because of increased capability and greater flexibility.

Category 6: Business Processes

Here the projects aim at improving the same or highly interrelated business processes. An example of this is in e-business implementation in which there are technical, marketing, and management subprojects that touch on several of the same processes such as customer service. It is very important to manage these projects as a group since otherwise work in one area can undo what was done in another project. This is an increasingly evident trend among projects since more companies are realizing that their strengths and competitive position are due to their business processes.

ASSESSMENT OF CURRENT PROJECTS

The following step-by-step approach will help you to assess your current projects as part of the transition to the new project management process.

Step 1: Gather Information across Projects

Collect the following information for all projects:

- Project plan in software file form
- Issues that are active, resolved, and potential
- Resources currently employed on the project
- Status of the project in terms of schedule and actual expenses versus the plan

Unfortunately, information will be missing because different project leaders have different styles. Information is also likely to be inconsistent. Some projects may not have resources assigned. Different projects may have different levels of detail. Don't let this deter you. Take what you get and move on to steps 2 through 4 in parallel. Don't keep asking people for more information. They might be embarrassed and defensive. You want them on your side.

Step 2: Establish Common Templates for Analysis

Begin by building a resource list. This will be a list of resources including staff, hardware, software, network, vendor, and business unit resources. Identify people by initials as well as by generic resources such as programmer and analyst. You will be using the generic resources in the template. When people turn a template into a project plan, they will not only add detail but will replace the generic resources with actual resources. Circulate the resource list among the project leaders.

While you are getting feedback, divide the projects into groups. The titles for Chapters 13 through 16 give you four categories. Build the templates from the top down and include at least three levels of outlining. (Templates are provided in the later chapters.) Enter the template into the project management software and circulate it as well.

After getting feedback, refine the template and resource list and submit it for another round of reviews. Next comes the detailed work. You must take one of the project plans and fit it into the template. Work with the project manager to do this. Hold a meeting to review this plan with the other project managers. The other projects can now be placed into the templates. You are likely to encounter resistance. If this occurs, then visit these project leaders and work with them one on one to develop their schedules.

What is a sign of success from this effort? The project leaders drop their old schedules and begin to update the schedule that you developed using the templates. Make sure that you store the schedules on a network server so that people can view each other's schedules (but can make changes only to their own schedules).

Step 3: Identify Interdependencies

From the discussion earlier, examine the plans for dependencies. A guideline here is to start from scratch and define these rather than using the existing plans. You can then validate the ones you have identified with the existing plans. How do you insert a dependency? If the templates are not established, then insert tasks into the existing task list that indicates the link. This will be a temporary manual link. If the templates are set up, then you can link the schedules if the project management software allows for this. Also, all should be simple tail-to-head dependencies. Don't try to get cute or exotic with lagging or leading dependencies now.

Test what you have done by extracting all dependent tasks and milestones into a new plan that includes only these dependent elements. From what you know of the schedules, are these complete? Are these logical? If these tasks were performed successfully, would all risk associated with interfaces disappear? You will employ this schedule to analyze the effects of schedule changes. Make sure that you maintain links to the original plans so that when their tasks are updated, the plan of dependencies will automatically be updated. Another guideline is to label each dependency by type according to the categories listed earlier. This will aid your analysis later.

Step 4: Define Common Issues

An issue can be a known problem or opportunity. Don't spend time sorting and comparing these here. Instead, take the issues as they are given to you. For each issue define the following characteristics:

- Descriptive information about the issues
- Related projects and task areas
- Impact and effects of the issue if it is not addressed
- Urgency and priority in solving the issue
- Status of the issue

More detailed data elements will be given in Chapter 7.

Step 5: Group Projects into Categories

Using the templates from step 2, the dependencies from step 3, and the grouping of issues in step 4, you can now try grouping projects in alternative ways. Try different groupings based on technology, business area impacted, common issues, dependencies, and methods and tools employed. Why group the projects? Because it will help you later when you seek to analyze the data from the different projects. You will want to look at different groups of projects in making your analysis.

Step 6: Analyze the Issues

You are now ready to analyze the data. Here are some analysis steps:

- Combine issues that are variations of each other.
- Sort all issues by severity and analyze the highest priority issues.
- Sort all issues by business units and analyze the ones for the business units with the most issues.
- Sort all issues by age and severity and analyze the oldest, outstanding major issues.

- Based on the sorting, group the issues so that you can search for a solution to multiple issues at one time.

Specific issues are discussed in Part IV. The general analysis approach is to attempt to answer the following questions:

- Why has the issue remained unresolved?
- Which issues require immediate and direct action?
- What problems do business units face?
- Which projects have the most unresolved, serious issues?

Step 7: Analyze Schedules and Resources

There is a useful way to organize the analysis. Divide it into four steps: identification and extraction of the information, visualization of the information on the PC or paper, analysis, and actions that follow from the analysis. Let's consider several commonly encountered examples.

- **Assessing a resource across multiple projects.**
  Let's suppose that you have three projects that share a common resource. You are concerned about how this resource can be allocated to the three projects.
  —*Extraction.* Filter or extract all high-level tasks of the template along with the tasks that involve the resource for each schedule. Combine the results from each schedule into a new schedule.
  —*Visualization.* Split the screen with a GANTT chart on the top and a graph or table showing the resource use on the bottom.
  —*Analysis.* You can now adjust the schedule and move tasks in the GANTT chart. The impact on resource usage will be seen immediately on the bottom as the bar in the GANTT chart moves. The analysis can be conducted in a meeting with the project leaders using a projector for the PC image on a screen.
  —*Actions.* Once you have an acceptable allocation and change, then the original schedules that use the resource can be modified to what was agreed to in the analysis.

- **Resolving issues across multiple projects.**
  The situation here is that you want to analyze and resolve one or more issues that are involved with several projects.
  —*Extraction.* Filter all of the projects involved with the tasks that are directly impacted by the issues along with the following, dependent summary tasks. This will allow you to see the impact of changes to the tasks involving the issues. Combine the results into a new schedule.
  —*Visualization.* You might again employ a split screen. The GANTT chart for the new schedule can be on top; on the bottom is a resource table that shows the number of hours each resource is used per unit of time.

—*Analysis*. As before, you can adjust the GANTT chart or change the tasks and determine the effect on resource consumption on the bottom.

—*Actions*. The most likely action is that after considering several alternatives, you will find a way of allocation and division of tasks that resolves the issues with minimal schedule impact overall.

- **Analyzing dependencies between projects.**
  Assume that you are considering a number of projects with different types of dependencies.

  —*Extraction*. Extract all major milestones and summary tasks from the schedules along with the dependent tasks. Combine the results into a new schedule.

  —*Visualization*. Split the screen with a GANTT chart for the extracted information and a GANTT chart for one of the projects about which you are concerned on the bottom.

  —*Analysis*. You can change dates and dependencies of the tasks in either GANTT chart and see the effects in the other chart. Doing this allows you to assess the impact of slippage. You can also modify and add tasks to change the dependency and see what happens.

  —*Actions*. This analysis can be used to narrow the scope of the dependency. It can also be used to assign higher priorities to one project so as not to impact others.

PLANNING ACROSS MULTIPLE PROJECTS

The preceding steps prepare you to plan over the longer term. That is, with resources known and the templates established, you are prepared to create multiple projects for each project idea. You can then combine these projects into an overall project. This summary project contains both ongoing work and potential projects. This analysis is critical if your organization is contemplating adopting a new direction in technology. Examples of the impact on resources might involve data warehousing, client-server systems, electronic commerce, wholesale replacement of legacy systems, and intranet systems.

SETTING PRIORITIES AND THE PROJECT SLATE

Several times a year, the management steering committee grapples with allocating resources: hardware, software, network, and personnel resources. For hardware, this includes not only what and how much will be acquired but also where it will be deployed. Staff resources that are being apportioned include operations and network support, systems analysts and designers, programmers, support staff, business unit staff, and contractor and consultant staff.

Types of Projects

You will have many projects and will have to divide them by type. Here is a set of categories that has proven itself over time. Assume that you have a long-range strategic-systems plan that has identified new project opportunities.

- *Operations support.* This is the daily work that keeps the software going. It includes backup, recovery, restart, and other programming activities. It also includes keeping the network and computers operational. Upgrades to operating systems and other related work are included here. This support is going to consume a great deal of resources.
- *Ongoing development projects.* Typically, there are some development projects that are continuing from the current year and will still be going a few months from the present time.
- *Architecture-related work.* This can include upgrades to hardware, system software, and the network. It can also include replacements. The work goes beyond installation. It includes testing, integration, documentation, and training.
- *Backlog of project requests.* From previous years there is typically a backlog of work that did not get attention last year. These items now present themselves.
- *Strategic systems plan action items.* These are project ideas from the long-range, strategic systems plan.
- *Maintenance and enhancement work.* Some of this work is often mandated by law or regulation. Maintenance is defined here as fixing problems and ensuring that the software continues to meet its requirements. Enhancements include the addition of new features and capabilities or the extension of capabilities.
- *New project requests.* These are new project ideas from divisions and the systems organization in addition to those proposed in the strategic systems plan. Some of these may be targets of opportunity.

Note that if you try to combine any of these, you could run into trouble since each area has its own criteria for evaluation. To free up resources you might want to determine if you can cut down the resources necessary for each area. Here are some comments by category:

- Could you ever reduce operations support? The answer is yes if you can simplify the architecture and modernize it.
- How can you control ongoing development resources? Determine how much of this is real development and whether sufficient progress is being made.
- How can you control architecture-related work? The best answer is through careful planning and staging of the work to be efficient. You might also get vendor resources.

- If you examine the backlog, you often find that the requirement or pressure for the project dissipated. The users were able to do something else.
- For the strategic plan elements you can take out the action items that do not require systems and technology resources. You thereby reduce the candidates exclusively to projects.
- For maintenance, go through each area and kill off the requests that lack benefits. The same applies to enhancements.
- For the other new project ideas, you really have to ask why these did not arise during the planning stage.

Some comments on analyzing and handling project requests are as follows:

- The need for the project is important. A government reporting requirement, for example, may be dictated. This can be stated here. The need for other types of items must be based on the business process and not the organization. It is often in the improvement to the business process where the benefits will be.
- For resources required, please identify the extent of scarce systems staff time needed as well as other resources. This will identify conflict situations between resources later and assist in their resolution.
- The impact, if not approved, is a very good item to require. It encourages people to conceive of a backup plan. They also have to discuss what they do now and how they have gotten along so far without the project.
- Divide tangible and intangible benefits. Over the years people have become more insistent on hard, tangible benefits. Many technologies and projects promise a lot, but few truly deliver. However, you can also acknowledge that real results often fall short of goals and expectations. Tangible benefits focus on cost savings or avoidance as well as revenue generation. For example, groupware might seem fuzzy and intangible, yet it speeds up a project. Speeding up a project lowers its cost, producing tangible benefits. Increased quality, improved customer satisfaction, and reduced turnover of staff are tangible. Ease of use can result in training savings.

How the project will be managed refers to the role of the business divisions and information systems. This item is important due to the number of projects that lurch out of control after starting because no one stepped up and assumed responsibility. If the information systems unit is to run all aspects of the project, then the business will not be involved. Lack of business involvement will likely doom the project.

After the initial review, classify project requests according to the categories covered earlier. All requests for new development or software acquisition as well as requests for enhancements and maintenance originate from business divisions. Operations support and architecture-related work come from information systems. Strategic systems planning project ideas are to be put forward by business units and not the planners. The business has to demonstrate support at this time. The exception here is the set of architectural projects that require the technical support of the information systems organization.

Before and After Allocation

Without a proactive process, resources are typically consumed by operations support, maintenance and enhancements, current development, and a few new projects. You probably have observed this repeatedly in diverse organizations. If you consider what is actually performed over several years, you see that the mixture does not change very much. This is how an organization can spend vast sums in systems and get nothing in return.

In the new process, the project slate composition changes, but not radically. Why? Because operations support consumes so much resource. Use some of the techniques presented here to cut down maintenance and enhancement and some development. You would be able to do some architectural improvement and some strategic planning projects. By diverting some resources into architecture and the strategic plan, you work toward long-term benefits.

The criteria for evaluation is specific to each category.

- *Operations support.* The analysis begins with assessing whether the numbers are in line with the past. As an earlier suggestion indicated, you might go back to systems and ask what they could buy or do to reduce this by 10 to 20%. This then leaves room for emergency fixes and repairs—work that cannot be predicted.
- *Infrastructure support.* This includes hardware, system software, and network support. Here look for escalating costs and labor due to aging equipment and software. You might see if you could get competitive bids for some of this support.
- *Ongoing development projects.* The first criterion is to assess whether the business requirement that generated the project is still valid. This is necessary. Projects have gone to completion without anyone ever asking about whether the business needed them. The next criterion is to determine if these are on target with respect to the schedule.
- *Architecture-related work.* The criterion here is to support the strategy in the planning process. However, there are many ways to do this. Should you spend all of the money this year? Or should you hold some back for next year to take advantage of technological improvements? Usually it is best to stretch it out, if possible.
- *Backlog.* Much of the backlog can be dropped. Often, the systems group does not want to appear to be uncooperative so systems personnel politely tell the business unit that it is in the queue of work to be performed. The division goes away and opts for an alternative. Nothing is done. Members of the division become so mad that they don't tell systems to remove the

division from the queue. The information systems unit, on the other hand, leaves it in the queue because excessive demand is impressive, making the information systems unit feel wanted.

- *Strategic planning projects.* While these plans originate from divisions, you still have to be willing to compromise. Will you compromise to get the best projects approved? You may end up backing high-priority project ideas in divisions with high levels of management support.
- *Maintenance and enhancement.* Here is where you get out the budget ax. Look for areas in which a business department has received dedicated support at a certain manpower loading for years. Also, search for work that was generated by individual deals between an IT manager and the supervisor of a user area without a review of real benefits.
- *New projects.* Rank these in terms of benefits. The new projects must be consistent with the framework of the plan and the architecture.

Each area has its own criteria. When you add them up, there is overcommitment in some areas; other areas may have spare resources. This is the fodder for trade-offs. The next analysis step is to assess the projects between categories or consider the pool in total. Try more than one approach.

Allocation Percentages

Consider specifying a percentage for each category that will provide limited support for new projects. Here is an example:

- Operations support: 20%
- Infrastructure support: 15%
- Ongoing development: 10%
- Architecture-related work: 7–8%
- Backlog: 5%
- Strategic planning projects: 5–8%
- Maintenance and enhancement: 25%
- New development: 10%

Are you surprised by these numbers? The proactive allocation leads to a gradual shift. Eventually, planning items diminish as does operations support and maintenance and enhancement. They are replaced by new development. Note that old systems and architecture are being addressed in these percentages. You are still performing substantial maintenance. After setting the percentages, apply them within each category.

Allocation Based on Fairness

Associate projects and work with business units. Assign a fixed percentage to enterprise-wide support (say 45%). Then allocate the remainder (55%) across the business divisions based on each division's importance to the business plan.

Allocation Based on the Strategic Systems Plan

Assign priorities based on the objectives and strategies within the strategic systems plan. The criterion might be alignment with the objectives.

Now review the results of each of the three items. You can even create a table. The rows are the projects and the columns are the allocation methods. In the table place a yes or no as to whether the allocation method included the item. Create a fourth column that contains an average of the scores or ratings.

A potential agenda for the meeting of the steering committee is as follows:

- *Introduction.* This section discusses the process followed.
- *Summary of the strategic systems plan.* This section provides perspective and a reference point for evaluation.
- *Summary of projects that made the first cut.* This section highlights those that clearly need to be supported. This agenda item ends with pie charts of the resources consumed and those still available.
- *Summary of dropped projects.* This discussion points to some of the projects that did not make the cut but are still viable candidates.
- *Discussion and setting of the slate.*

As a result of the meeting, you will have the slate along with some projects to which management specifically assigns high priority. Other projects will be waiting in the wings for resources.

A suggested outline of the project slate report is as follows:

- *Introduction.* Describe the process followed.
- *The project slate.* Identify each project in terms of description, schedule, project leader, costs, and benefits.
- *The projects that almost made the slate and that may be activated during the year if resources are available.* Again, provide description, costs, and benefits.
- *Summary of the state of systems and technology if all of the project slate items are completed.*

For this last item, try a scenario. That is, consider several business processes. Give the before and after work flow in the processes. This is also useful for highlighting the interdependence of the projects and for planning action items.

EXAMPLE: ASTRO BANK

Astro Bank is a leading consumer and commercial bank that operates in multiple states. The bank had relied on many old legacy systems. When it was decided to modernize these systems, management insisted that the same programmers doing maintenance also develop the client-server and distributed systems. The result was a disaster. Maintenance suffered and no new system resulted. The approach was abandoned and a new distributed systems group was established for the new development. The strategy initially was to develop new systems that employed the legacy systems for batch processing. After these new systems were developed, the legacy systems were to be replaced or modernized. At the start of the analysis, the two groups were not cooperating with each other. The legacy systems staff continued to do enhancements and failed to inform the distributed systems group of what they had done. The software on the distributed systems side blew up. The distributed systems staff in frustration began to build modules and functions into the new software that replicated the functions in the legacy system. It looked like another disaster coming down the track.

In the first step, information was gathered from each group separately. Each used different project management software, different methodologies, and entirely different management styles for resolving issues. Neither side really took advantage of lessons learned. Building the templates had to be conducted separately with each group (step 2). A new resource list was created. Most tasks had to be edited into a common framework that was between the styles of the two groups. The interdependencies were difficult to establish because neither group had tasks that related to the other group due to a lack of communication (step 3).

Issues were also a problem (step 4). The distributed group provided a list of issues. The legacy group did not acknowledge these issues. The legacy group provided user requests for maintenance and enhancement. Additional interviews were required to solicit the information to build the issues. Several methods were used to group projects (step 5); they were grouped by business unit, by legacy system, by network, by issues, and by banking product. Each of these categories provided different insight into the projects.

The analysis of the issues (step 6) revealed that many issues were missing. This was reinforced from step 3 when many new tasks of interdependence had to be defined. The new issues related to these tasks. A meeting of project managers was held as part of step 7. The order of presentation was as follows:

- *Review of the revised schedules of each group.* This showed a common structure and introduced templates that crossed both groups.
- *Review of the issues from each group.* This revealed a common set of issues.
- *New tasks and issues.* These were presented along with the impact of what would have happened had these tasks and issues not been defined.
- *Priority list of issues.* This triggered a discussion of the issues that were faced by both groups.

It would have been asking too much for the two groups to cooperate immediately and fully. The cooperation built over time. The Astro Bank example will be followed up in later chapters.

E-BUSINESS LESSONS LEARNED

Managing e-business implementation often means managing a set of related projects. It is a mini project slate in which you struggle to allocate people across subprojects as well as work with the dependencies between milestones and tasks across the subprojects. Here are some examples of dependencies.

- Marketing needs to specify the discount structure for merchandise so that both IT and business departments can handle the discounts as soon as they are offered. The same applies to promotions.
- Existing procedures and policies within a department may have to be modified to be synchronized with those imposed by e-business. Otherwise, the customer or suppliers on the less favorable side will begin complaining— giving rise to a crisis.
- The same people are needed in different subprojects because of their expertise.

The situation is even more complex when you move up to the overall organization. E-business implementation requires detailed knowledge of the company's business rules. Unfortunately, this information often rests in the minds of a few critical and senior employees. These people are required for normal work—just to keep the business going. How will you allocate their time between normal work and the e-busienss subprojects? That is a challenge. In one case, the company dedicated the people to the e-business effort and the company experienced revenue loss and customer service complaints. In another, a firm gave priority to normal business. It took forever to implement e-business and then it did not work right. There must be a balance.

GUIDELINES

- Should you create multiple systems projects in a single business unit? Only if the projects are directed at the same goal. Otherwise, the projects can tear the business department apart by requiring extensive time and involvement by key business department staff.
- With success, a single project that is not finished can spawn other projects. This occurs because management sees success and wants to spread it around. It is a natural reaction. The problem here is that these additional projects will likely require some of the key staff from the original project. The original project can be delayed and put in peril.
- Treating all projects the same can smother smaller projects. The approach here allows for smaller projects to be tracked within the template and issues

as larger projects. There is no increased overhead in management or administration unless there is risk in a smaller project. Risk translates into more issues and a greater attention to the schedule. This is really what you want—the degree of attention and management effort depends on the risk, not the size of the project.

- Related to the preceding guideline, you need to treat small projects as projects since if they just exist as work, there is a lack of control.
- Try to avoid long-term resource allocation. Systems projects can change in terms of schedule quickly. Therefore, experience indicates that doing long-range allocation of scarce resources across projects is only useful for general estimation. Allocating far in advance can lead to inflexibility if you have to shift resources later. It makes sense to analyze and allocate resources for a month or two in advance. This can be done in a rolling mode where each week a new week is added.
- Establish customized data and formats for the project management software in advance. If you attempt to learn the details of the software at the same time you are gathering the information in the early steps mentioned previously, you will probably not do a good job with either. Learn the software and establish any customization up front.

WHAT TO DO NEXT

1. For each category of interdependence, identify several projects. Next, assess how these dependencies are currently being managed. This will indicate some of the benefits to managing the portfolio.
2. Collect the existing project plans and begin to build a resource list that can be employed by all of the projects. Then start to build high-level templates. To get support from project leaders, solicit feedback on an iterative basis as you go.
3. Perform a dry-run analysis to see what resources could be moved between projects. This will be a sensitive topic, so you present results in the aggregate. In other words, state that if projects were managed collectively, 10 to 15% of the resources could be reallocated. Don't name specific projects.

SUMMARY

From a business perspective, systems projects are highly interdependent. From a systems perspective, it is necessary to divide them into single projects for purposes of management and accountability. It is a dilemma. To get your cake and eat it too requires you to impose standards on all projects so that you can perform analysis, trade-offs, and allocations across projects while at the same time preserving each project's own individuality and integrity. It is a delicate balancing act. However, experience shows that the benefits far outweigh the effort.

- Lessons learned can be applied across many projects, thereby improving work quality and performance.
- Resources can be shared among projects to obtain some economies of scale.
- Information systems can take advantage of new technology across multiple business processes and systems faster.
- Issues that once were submerged in single projects can be surfaced.
- The project management process is more consistent.
- Information on projects is shared among systems and business managers and staff, leading to greater cooperation.

Developing Your Project Plans

Chapter 4

The Project Concept

INTRODUCTION

The first three chapters laid out a project management process and provided suggestions for managing multiple projects. The detail now begins. Part II starts with the initial concept of a project and ends with the completion and approval of the project plan. From there we will move on to the management of the project and address specific types of systems and technology projects.

In the past someone came up with an idea for a systems project. This quickly led to the development of a project plan. The budget and the schedule were then derived from the project plan. Systems projects are often notoriously vague when they are started. This fact, combined with the rapid development of a plan, can produce many problems later. Experience and lessons learned point to the benefits of hashing out the purpose, scope, and other factors in the project first—prior to developing the plan. If you wait until a plan is in place, there is so much detail that most managers will not question the plan—or the detail.

A project concept defines the purpose, scope, roles, benefits, issues, and framework for a project idea. The notion of developing a project concept has demonstrated a number of benefits, including the following:

- Management can perform trade-offs at a high level relative to purpose and scope. Specifially, the extent to which reengineering and architectural improvement will be part of the project can be set.
- The business unit roles and responsibilities can be defined. Whether the business unit will step up to these in the project can be validated.
- Related to the roles and responsibilities, the benefits of the project to the business can be roughly estimated.

- Because management and business units have an early role and involvement in setting objectives and scope, there is less likelihood of changes later in the project.
- Key interfaces with other systems and technology can be determined.
- It is less expensive economically and politically to address issues at this time than to wait until the project has started.
- Management gains an understanding of the issues and potential problems that will potentially have to be faced in the project early. These can be discussed rationally before the issues arise—an early warning system.

There is an analogy between the life cycle of systems and this approach for projects. The project concept corresponds to the feasibility stage of a project. The development of the plan can be seen as requirements and design.

Developing a project concept is valuable for any project management process. The project concept allows you to do an organized analysis of the project idea in the process. It also helps you build management support prior to developing a detailed project plan. This chapter deals with the development of a project concept. After completing this step you will have management approval for the project team to be identified and the project plan to be developed.

APPROACH

WHAT IS IN THE PROJECT CONCEPT?

Here are the key ingredients:

- *Objectives.* The objectives of the project are set. Objectives here mean the impact on the business processes or technology base. Pinning down business impacts as opposed to just completion of the setup of a system will yield a better estimate of system benefits.
- *Scope.* What is in and what is out of the project? Experience reveals that although many can agree on the objectives of a project, much disagreement can arise as to scope. Considerable time will be spent here.
- *Roles of IT, business units, and vendors.* What activities in the project will each party be responsible for? Will they participate in evaluation, design, installation, training, and so on? Will they help manage the project? How will issues get resolved? How will the parties interact with each other?
- *Major milestones and task areas.* Using the template you can develop rough estimates of the work ahead and milestones that have to be met.
- *Benefits, schedule, and costs.* With the objectives, scope, milestones, and tasks, you can estimate the benefits, overall schedule, and costs. These will not be precise, but they will help management to understand what it will take.

- *Initial issues.* What are known issues at this stage? There can be resource shortages, business policy questions, and resistance to the project, to mention a few. Identifying these issues at the start prepares everyone for the road ahead. If you use an issues database and checklist, then you can identify which issues are applicable to the project.

Do you just write down what comes to mind and go ahead? No, this is an opportunity to follow an organized approach and prepare some alternatives. Management input can then be sought as to what is most suitable. An interesting behavioral phenomenon occurs. Many systems people start with a narrow purpose and restricted scope. When the concept is completed, it often is more broad and includes more of the business activities as well as technological support. It is important that as many managers as possible be involved in reviewing and discussing the project concept. In that way you are more likely to gain consensus and have fewer problems later.

CONCEIVING OF A PROJECT

This is a good time to be creative. Before you start nailing down the concept, you can consider a wide range of alternatives to address the situation. You may want to avoid doing a project and address the situation some other way. Each project stems from a combination of business and technical needs. Here are some potential alternative strategies.

- **Do nothing.**
  This is important because it indicates what will happen if the project is dropped. What will really happen if no project is done? Here are some detailed questions:
  —Will business units go out on their own to get other systems?
  —What will be the additional costs if the system and changes are not done?
  —What is the impact on the competitive position of the organization? How price competitive is the company without additional automation?
  —What are competitors doing?
  —Will the price or performance of the technology really improve by waiting?
  —If the company does wait, then when is the logical time to resurrect the project idea in the future?

- **Move work to another department.**
  Under this alternative, the work that the project targets is moved to another department. The logic here is that the problems in the current situation stem from problems within the department where the work is being performed.
  —Which departments can handle the work?
  —What problems in the current process can be related directly to the people?
  —Are there training and morale problems that could be addressed by this change?

- **Outsource the work.**
  Instead of fixing the problems yourself, you can think about having an outside firm who is more familiar with the process do the work for you. This is typically considered for many support functions. Consider these questions:
  — Is there a competitive advantage in being able to market now that you are not doing the work yourself?
  — Is there internal, company-sensitive information that cannot be allowed to go outside?
  — Is there any special knowledge that the internal staff must have to do the work?
  — If the work were to be outsourced, what is your firm's dependence on the outsourcing firm?
  — What efficiencies does the outsourcing firm bring to the table that your organization lacks?
  — What might be the cost of outsourcing?

- **Change the business process, not the system.**
  With this alternative, the procedures of the business process are updated. Additional training in the business process can be done.
  — What is the current state of the business process? Are the problems related to systems and technology?
  — What is the staff turnover in the department?

- **Change the system, not the business process.**
  This is the narrow definition of the scope of a project. The key question is, if you plug in a new system and keep the process the same, will there be any benefits? This can occur if the current business process is efficient. However, this is unlikely in most cases because people over time tend to build the procedures and handle exceptions around and through the system. If the new system did not address the changes and exceptions, then it might not be very effective.

- **Automate less.**
  Under this alternative, you consider removing some of the automated system from use and then simplifying what is left. Why is this a viable option? Because the business could have changed after the system was implemented. The old system does not fit well with the new process. Also, the new process may not be stable so that implementing a new system may yield more problems and no benefits.

- **Cut down on the number of staff members involved in the business process.**
  The concept here is not to change any procedures or systems. Instead, you consider reducing the number of people involved in the work. This is a possibility if the overall workload dropped but the staff was not downsized.

- **Increase the number of staff members involved in the business process.**
  Under this alternative, people are added to the process. This might be a short-term solution if the project was not approved. Adding people to handle work is a traditional solution. Consider the following questions:
  — If you added more people, how long might it take for them to be effective?
  — What is the cost of the additional people, facilities, and support?
  — What additional systems support is required for the additional people?
  — What happens to the error rate and rework?

- **Split work among departments.**
  The approach here is to make the existing business process disappear. Functions are divided and assigned to different departments. This strategy is employed when companies are merged and reorganized. The thinking behind it is that if several departments already do similar work, then some could take on additional work with minimal impact.

- **Change business policies only.**
  This alternative requires some discussion. Business policies are the basic rules for defining how the business unit will perform a business process. If you change policies, you can impact the workload dramatically. Suppose that your firm had the practice of issuing letters to new customers and you are considering a small project to automate the letter generation. What if you decided to not send out the letters because you could find no discernible benefit? This change of policy eliminates the need for a new system.

- **Change the business unit organization.**
  The concept here is that the current organization and management are getting in the way of the process being performed. If new management were put in place, then the performance of the business process might be improved.

Consider creating a table such as the one that follows. The first column lists the alternatives. The second column examines how implementation might proceed. The impacts of the alternative and benefits are the subject of the third column.

Alternative	Implementation	Impacts and benefits

Considering these alternatives is a good idea because you can later explain to managers and the project team all of the various alternatives you considered. You did not just plunge in and go for automation because you are familiar with information systems. Let's put it another way. If you don't consider these alternatives, management may ask you to do so later. This will harm your credibility and erode support for the project.

The approach for developing the concept is broken down into seven steps. Because client-server and other major examples are addressed later, several other examples will be considered here. One is the decision to upgrade PC software for office functions and electronic mail. This seemingly straightforward project can be a nightmare. The second project consists of implementing imaging for an insurance company, such as Beaumont Insurance.

STEP 1: DEFINE ALTERNATIVES FOR THE PROJECT OBJECTIVES

A suggestion is that you begin with a very narrow systems objective. This might typically occur in the installation and completion of a system. Nothing is mentioned here about business impacts or change. In the two examples, this might be stated this way:

- *Example 1:* Upgrade the software and hardware for the employees' PCs.
- *Example 2:* Install and test the imaging hardware and software.

These are traditional, narrow systems goals reminiscent of past decades. Now look under the sheets. In the first example, the staff will be upset and disturbed because the software that they use most and are most familiar with is being replaced. Their productivity is affected. How will all of these people be trained? How will their current files be converted to the new software? If there are letter and other word processing and spreadsheet templates, will these be recreated on the new software? Who can the employees call for help and questions? It is clear that the purpose must be expanded if you are to ensure success. If you don't do this, then some of the questions will not be addressed. They will fall through the cracks.

For the second example just installing imaging does not yield any business benefit. Forms have to be programmed so that you can employ optical character recognition (OCR). The image software must link to the software applications and databases. Suppose, for example, that you have loan application processing and servicing in mind. Then you scan the loan application form and have the system fill a data entry screen with what it recognized. The system should route the entered application to someone who can determine the applicable rate and whether you will issue a policy. These steps constitute much more work than just installing hardware and software.

In each example the narrow objective is just the tip of the iceberg. It is the first phase of the project. It is not the project. Here are some expanded objectives.

- *Example 1:* Implement new end user software and ensure that it is in use.
- *Example 2:* Implement a complete image system that front-ends the existing application processing system.

These are better. They are more complete. Notice that there is much more to each objective. You want to keep the wording short. However, there are still problems with these statements. No expectation or benefit is identified. If you were a higher-level manager, you might ask, "Why should we spend the money to do this project?" As a project leader, wouldn't you want everyone to keep in mind why people on the team are doing all of this work? Therefore, you must further expand the objectives to include the major benefits. Doing this, you can define the revised objectives:

- *Example 1:* Implement new end user software to provide new capabilities and increase employee productivity.
- *Example 2:* Implement a complete front-end system to increase efficiency and improve management control.

A note on the second example. Increased efficiency comes through reduced data entry and ability to route the transactions among different employees. Management control is improved because the system can track each transaction to determine how long it took to complete and the time spent in various states.

In forming objectives, here are some guidelines:

- Ensure that the objective is sufficiently broad as to include the business role.
- Include major benefits in the wording of the objective in business terms.
- Employ a standardized sentence structure starting with a verb and ending with the benefit.

What are the benefits of taking this approach? First, you provide a convenient way to list the various projects so that business managers understand them. Second, there is uniformity. Each new project idea has to clearly identify its benefits at the start. A third benefit is that by focusing on objectives in business terms, you are more likely to achieve benefits since the emphasis is on the business rather than implementing a system

STEP 2: DETERMINE ALTERNATIVES FOR THE SCOPE OF THE PROJECT

If you have the right objective, why is scope important? Scope determines what is in the project and what is out. Through the scope you will build the major headings of your project and template. Cut the scope and you cut the project. This is difficult for managers to see if they are looking at a task plan of 500 tasks or more.

A systems and technology project idea can be viewed in a number of dimensions: underlying technology, business unit involvement, extent of business process impact, extent of interfaces to existing systems, existing technology interfaces, impact on business policies, and impact on the business unit organization.

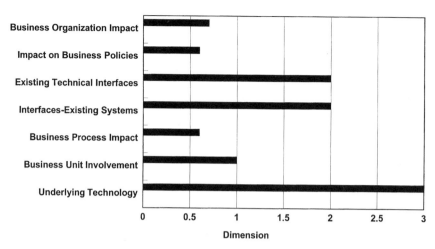

Figure 4.1

These dimensions allow you to compare different scopes. In example 1, if the scope of the project was just the systems group, then you have the chart shown in Figure 4.1. This chart shows that there is virtually no business involvement or impact. If you expand the scope to match the revised objective that was developed, you obtain Figure 4.2.

Example 2 is more complex. It was already agreed that the business process is involved. There are additional questions related to scope. When you change the work flow in the data capture process, you impact data entry and servicing. Therefore, it is natural to consider changing the organization. Because the new system can automate the business rules for insurance application approval, you might want to change business policies to allow automatic approval if the application meets certain criteria.

The first action in working with scope is to define alternatives and graph them as was done earlier. The second action is to determine the impact of different scope alternatives. The following table is suggested, using your information for example 2.

Scope	Business impact	Project impact
Imaging technology only	Few likely benefits	Narrow project with little risk
Imaging, work flow, linkage	Some benefits	Manageable, but complex
Work flow, organization, policy, etc.	Assured benefits	Project becomes political

As you expand the scope, you are more assured of the benefits, but you raise the level of risk as well. The graphs and tables are what you will present to management later.

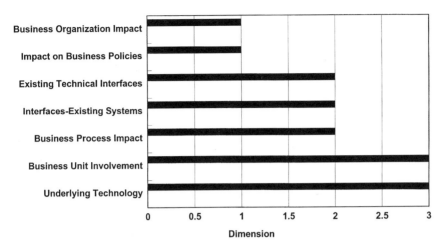

Figure 4.2

The third action is to employ the table and develop alternative project ideas. From example 2 it appears that the problem is how to deal with the organizational change. Certainly, this involves the business units rather than information systems. Here are some alternatives:

- *Alternative 1: Keep the entire project as one.* A problem with this approach is that a systems person is logically the project leader through the implementation of the system. Then a business person can take over to analyze and implement organization change.
- *Alternative 2: Split the project in two.* The first project ends with the use of the system. The second project is a business project dealing with business policies and organization.

How do you document the scope for an alternative? How about another table where the dimensions are rows and the second column indicates the scope. For the first project of the second alternative, you have the following:

Dimension	Comment
Underlying technology	Network, application software
Business unit involvement	Substantial, at a tactical level
Business process impact	Limited to procedures
Existing application system interfaces	Application processing system
Existing technology interfaces	Image software to front-end software
Business policies	Not applicable
Business organization	Not applicable

How does an objective relate to scope? You can have the same objective for several alternative scopes. The scope has to fit the objective. Because the objective is general, there is room for several alternative scopes, as was done with example 2.

You can also prepare a spider or radar chart indicating different alternatives for scope. The dimensions for the chart are as follows:

- *Business processes.* To what extent does the alternative involve the business process?
- *Benefits.* What business benefits result with the scope of the alternative?
- *Risk.* What are the project risks with the scope of the alternative?
- *Costs.* What is the magnitude of the cost of the project wit the scope of the alternative?
- *Elapsed time and schedule.* How long would the project take with the scope?
- *Systems and technology.* What is the degree of work and change required by the alternative?
- *Organization.* To what extent does the business unit have to be involved in the project to make it successful?
- *Policies.* To what extent must policies be addressed and changed with the scope?

You can now draw lines linking the appropriate levels for each dimension. This helps you trade off between different alternatives. Experience shows that it is the scope where management input and discussion are most valuable.

STEP 3: PREPARE ALTERNATIVE SCHEDULES, COSTS, AND BENEFITS

For several alternative purposes and scopes, you now will try to figure out costs and benefits by developing a general schedule. The temptation here is to plunge into detail. Don't succumb to this temptation. Remember that several alternatives are available.

The first action is to define a high-level list of tasks based on the narrowest scope. If a template is already available, then start with it. Otherwise, proceed from the top down and define the first level of tasks. Here is a cut for example 2.

1000	Define the detailed project plan.
2000	Define new work flow and process for data capture.
3000	Implement imaging hardware and software.
4000	Design and implemented the front-end system.
5000	Customize and interface imaging software to the front-end software.
6000	Implement interfaces to the legacy application system.
7000	Develop new procedures and train business unit staff.
8000	Implement new business process and system.

Some comments are appropriate here. First, each task is numbered so that it is easy to follow. Second, each task begins with a verb so as to indicate the action required. Third, the project ends with the new process in place. There are no policy or organization changes. To expand the scope, you can add these tasks:

9000	Analyze business policies and organization.
10000	Implement new organization and policies.

You now have two task lists at a very high level. Next, determine the dependencies between these tasks. This is your second action.

The third action is to identify resources for each task. Don't do this by name. Instead, indicate the type of person needed (for example, systems analyst or programmer) and the number required for each task. Later you will refine this by adding more tasks and more detailed resources.

The fourth action is to develop an estimate for how long each part will take. Be conservative. Base your estimate on previous project experience. Where you perceive risk, add more time. For example, you might think that task 9000 will take about a month. Don't kid yourself. It will take longer because human resources and management will have to determine which people will move where. Allow three months. The same is true for task 10000.

With the estimated schedule and resources, you can now perform the fifth action—estimate the cost of the project. Many people fail to identify certain types of costs. Be as complete as you can. If you require some people on a part-time basis, include them. If you need to test hardware and software, include a charge for the use or acquisition and use of this technology.

STEP 4: DEFINE THE ROLE FOR THE ALTERNATIVES

The role of the business unit flows from the objectives, scope, and schedule you have created. The wider the scope, the more likely it is that business unit role will expand. If the scope grows to provide additional systems functions, then this expands too. To illustrate this role you can develop another table. Here is a partial table for example 2.

Task area to be performed	Business unit doing work	Comments
Training of end users	Data entry and operations	Requires train the trainer
Development of procedures	Data entry	Needs guidance and support
Policies for automated approval	Management	Must match organizational model

What can this table be used for? Develop a table for each alternative that you considered in step 3. You are now in a position to discuss the impact of scope on benefits, schedule, and the business unit role. Expand the scope and you expand the role. Without this many managers opt for the wider scope. After all, it has greater benefits. However, the downside is the longer schedule and the expanded business unit role. Potential business unit roles include the following:

- Recurring training of employees in business unit procedures and systems use
- Training materials for the process and system
- Initial training in the new business process and procedures
- Operations procedures for both the process and systems
- Definition of the new business process and procedures
- Comparison of the new and existing process to get requirements and benefits
- Assessment of the benefits after the new process and system are put in place
- Participation in testing and conversion
- Analysis of the current business process and identification of issues in the process

Notice the order of this list. This is the order in which you should discuss the user role with the business. It gains commitment.

The same comments apply to both the IT and vendor roles in the project. As you define different alternatives, you can then assess how IT, business units, and vendors will interact and interface. If the business unit is unwilling or unable to assume the appropriate role, then this is a good time to stop work and not proceed with the project—unless they are committed at the start, history in projects reveals that it is more difficult to get the commitment later.

STEP 5: ANALYZE THE RESULTS OF STEPS 1–4 AND DETERMINE THE RECOMMENDED PROJECT CONCEPT

In this step you will perform any further analysis of the work you have done. You may still have several alternative project concepts. It is in this step that you contact some of the managers and let them review the material to get their input on the scope of the project. Start with a discussion of the schedule and the business unit role. Once this is understood, move to what happens when the scope is expanded. Use their input and feedback to modify and update the tables and estimates.

STEP 6: IDENTIFY INITIAL ISSUES

In this step you can now identify issues that you know will appear. These suggestions for example 2 are organized by the chapter headings in Part IV. Note that this is just a sample of more than 75 issues that were developed.

- **Business issues.**
    - —Can OCR be effective if only a small number of forms are programmed (or is this a situation in which automating 5% of the forms will give you 65% of the volume?)
    - —How will exceptions in the work flow be handled?
    - —Do other departments require access to the image?
    - —Can the paper be destroyed after scanning?

- **Human resource issues**
    - —How long will it take for the programmers to come up to speed in programming the image software?
    - —What will be the resistance of the business unit staff to the system?
    - —How will job descriptions be developed?

- **Technical issues.**
    - —What are the network requirements to support the routing of images? Can the current network support this workload with adequate response time?
    - —Can the application software be modified to accommodate a link to the images?

- **Vendor and consultant issues.**
    - —What support will the image vendor provide?
    - —What support will the vendor of the insurance application software provide?
    - —Are the tools provided by the different vendors compatible?

As a result of discussing issues, management gains a better understanding of the challenges that will have to be faced. The issue discussion may lead backward to revise earlier steps. In general, issues tend to bring people down to earth so that it is less likely that there will be problems in management and user expectations later. Also, when some of the issues later emerge, then there is less surprise.

STEP 7: OBTAIN MANAGEMENT FEEDBACK AND COMMITMENT

In this step you seek agreement to develop the project plan and more detailed budget. Here management has the opportunity to view the project concept without massive pressure to get the project underway. Remember that you seek agreement on objectives and scope and the other items. The approval of the resources and detailed plan will come later.

Assume that you have already obtained some management feedback from the previous steps. What are the detailed goals here?

- The business unit management must commit to the project.
- Information systems management must support the project in terms of making resources available and juggling priorities to support interfaces, the network, and so on.
- Upper management must see the benefits and resources required.

Business unit commitment is revealed in the roles defined in step 4. There is commitment from upper management in the unit as well as from middle-level managers and staff.

Here is a useful series of actions that you can take with respect to management presentations. The first step is to get information systems support for the project. You can use the issues that you have identified in the previous step to indicate the type of support that you require. If you fail here, then it may be because the information systems unit places a higher priority on other work. Management may feel the same way. The project concept is dead in the water.

Let's assume that you have passed this first hurdle. Turn to the business unit next. Approach the business unit managers that you have been in contact with and inform them that you desire to present the project concept to a wider audience in the business unit. You are attempting to reach the people in the departments that will be working with the new system. You want their support to buttress the support of their own management. They can also provide you with more details about benefits.

After you have lined up information systems and the business units, then approach individual upper-level managers either directly or through intermediaries. When you come into a management meeting for support, introduce the project and then turn it over to the business unit manager first. The business unit manager can indicate the benefits and the business unit's commitment to the project. The information systems manager can follow indicating that the approach is feasible, doable, and can be supported.

EXAMPLE: ASTRO BANK

The initial work started out badly. Recall that there were two systems groups. The distributed group plunged in and developed a project plan. It was too narrow. Not only did it leave out the legacy system tasks, but it neglected changes in business procedures, policies, and organization. The plan was rejected. The approach was to hold several planning sessions to determine the scope of the project. Most people agreed on the overall purpose. The problem initially was the scope. The application was an on-line servicing front- and back-end system to an installment lending legacy system. Some of the questions that arose were the following:

- Will changes be made to the legacy system since the legacy systems team was going to have to do work to support the interfaces to the on-line system?

- Will the business unit do loan servicing at its branches as was the case? Or will servicing be centralized?
- Is the telephone system to be modernized to support both incoming and outgoing calls on a more automated basis? What are customers able to do for themselves?

Answers to these questions determine the extent of benefits. If the business process, organization, and telephone system are left intact, then there are few benefits. However, both systems groups felt uncomfortable about expanding the scope to these areas. A second meeting was conducted with these managers and the business unit responsible for installment lending. In this meeting the scope was expanded to include all of the elements mentioned.

With the scope settled, the responsibilities of the business unit were addressed. At the start of this discussion, the business managers indicated that they had no staff or time to spare. They wanted the systems groups to do documentation, testing, and training. It was pointed out that (1) this would delay the implementation by months and (2) the support provided by systems in these areas would not be business specific. Reluctantly, the business unit manager committed to the role.

At the next meeting, the purpose, scope, and roles were reviewed. Attention then drifted to the schedule and benefits. The business unit manager was reticent to sign up for specific benefits. To get down to details and resolve the issue, the outline of the new business process for servicing in a centralized mode was developed. This led to a comparison with the current process and the identification of the saved steps and labor. The business unit was assigned to then define the benefits.

When the schedule was presented, major problems erupted over sequencing of the project. Will the new system be implemented first? Will the telephone system be installed? Or can the staff and servicing be centralized? These were the most important questions after the scope. In the end it was agreed to centralize first. This might standardize procedures and allow for the organization to become stable. Then the telephone system can be installed. Finally, with the organization and communications in place, the system can be implemented.

A later meeting was devoted to defining the initial list of issues. More than 100 business, technical, and policy issues were defined. Some of the following issues were raised:

- What will happen to the staff in the branches?
- What access does the branch need to loan data if there were a central servicing area?
- What was the bank's policy on the role of the branch staff?
- Do any other business unit staff members require access to the new system beyond those in the servicing center?
- How will the transition at each stage be done?

- Who will serve as the project leaders?
- What will happen to the backlog of user requests for changes to the install-ment loan legacy system?

Note that if the scope had been narrowed to the system, most of these issues would disappear. The network would have had to be extended to hundreds of branches to support servicing.

E-BUSINESS LESSONS LEARNED

The project concept is essential in e-business projects. Management typically sees e-business implementation as an IT project. It is not. IT work is quite limited compared to the overall scope of the effort required. Some of the elements that typically have to be addressed are the following:

- Changes in marketing strategy and policies
- Definition of sales, promotions, and discounts
- Changes to existing business processes to make them synchronized with e-business
- Organization change to support e-business
- Definition of new business processes to support e-business

Suppose, for example, that you are a retailer moving to the Web. You would have to handle orders, customer service, returns, refunds, discounts, billing, and other transactions differently to support e-business in many cases. The techniques used in the preceding steps turn out to be very valuable in determining the imple-mentation strategy. Some additional comments are as follows:

- By discussing alternative objectives, you can mesh the implementation with the overall strategy of e-business for the firm.
- Through comparing alternatives for scope, yuo can see how much will have to be changed.
- The issues analysis paves the way for a more realistic assessment of the schedule.

GUIDELINES

- When defining an objective, if the wording of the objective is too complex, then consider breaking it into several projects. For example, suppose your objective was, "Implement a client-server system and wide area network so as to improve our competitive position." You can simplify the objective by establishing the client-server system and the wide area network as separate projects.

- The objectives and the scope do not have to match. The key is that the objective be covered within the scope. The scope gives you room to work within, while the objective provides focus.
- When expanding the scope of the project idea, remember that often you will have to extend the scope in several dimensions to be consistent. For example, if you wish to include the manual steps of the business process in a project, then you are likely going to want to include the other systems and PC-based tools used in the business unit.
- After defining potential issues in each of the preceding categories, you can rate them according to different criteria:
  —Importance to the project overall
  —Importance at the start of the project
  —Impact if the issue is not addressed
  —Organizations, technology, business process, policy, and so on involved in the issue.
- To show how important these issues are, select one issue of importance early in the project, develop alternative solutions for the issue, and then define the impacts on the project for each solution.

WHAT TO DO NEXT

1. For a creative exercise, consider looking around for a potential small project. Examine this potential project in terms of the alternatives defined in the chapter. Develop the table that was presented.
2. Take several projects and see if you can formulate the objectives of each project.
3. Consider one project that is not meeting its schedule. Define the purpose and scope as they were at the start of the project. Now do the same thing for the current state of the project.
4. For a project with which you are familiar, develop the scope in terms of the charts and tables identified in step 2.
5. Drawing on the previous exercises, define the business unit involvement that is both possible and feasible. What types of resistance do you think that you might encounter?
6. Develop an initial list of issues in the categories defined earlier.

SUMMARY

This chapter helped you to develop the concept of a project prior to developing a plan. Even if a new project management process that recognizes the project concept is not in place, it is very useful to think through the steps presented in this chapter. Doing so will solidify your support for the project. Another benefit is that a project concept provides a framework for developing the project plan later as well as for enlisting people to be on the project team. Developing the project concept can save you time and avoid issues later. The project concept allows management to put the objectives, scope, roles, and other elements into clear focus.

Chapter 5

The Right Project Leader

INTRODUCTION

With larger and more complex IT projects, companies face formidable challenges in attracting, managing, and supporting project leaders. Some of the questions they must answer include the following:

- How will new project leaders be found?
- Will there be a project office to direct all projects?
- What autonomy will a project leader have?
- How will project leaders be trained?
- How will we cope with project leader turnover?
- How will we deal with a shortage of project leaders when the projects must still continue?
- How will analysis be performed across projects?

These are but a few of the questions many firms have cited in surveys. The purpose of this chapter is to examine the role of the modern systems project leader. This is different from traditional approaches in that the role of the project leader in systems and technology projects is different today from what is was 10 years ago. Today, there are more issues involving business units, processes, and external factors in addition to new technology.

The scope of the chapter is wider than the traditional approach that concentrated on duties and gave too much attention to the mechanics of project planning. The duties of the project leader also change within the project as implementation is finally reached. For example, for some large projects, it may be useful to have a separate project leader for the implementation and rollout of a system to a large business unit.

To highlight the differences between traditional and modern, let's examine the traditional project manager. Individuals in information systems started out in a technical role. They might have been systems analysts. The systems manager might select one to be the manager of a small project. Preparation and training was limited. Often, people were just thrown into the situation. They might have been offered some training in project management. This training typically focused on project management concepts: GANTT, PERT, CPM, and so on. Each project manager was on his or her own. Project managers reported to the IT manager and resolved problems one-on-one with the manager. The traditional approach worked because projects were mostly independent of each other. The business unit had little dependence or involvement. Life was simpler. When a large project came along, the same method was used. This method of managing project leaders had a number of shortcomings that persist today. Look at the following list to see how many apply to your firm.

- There are no common schedules or approaches across projects because each manager is on his or her own—making it difficult to roll up data across projects.
- There is no sharing of information. Sharing of information might be seen as a sign of weakness.
- Only mechanical-technical training is available for project managers.
- There is a lack of coordination on solving problems among projects. Problems are solved one at a time on an ad hoc basis.
- There are no standard approaches for reporting on projects. Each project manager can determine how to report.
- Project managers use a different variety of software tools.
- No lessons are learned from projects.

APPROACH

To make this discussion more useful and interesting, project manager topics have been organized around key questions that you would ask.

WHAT SHOULD YOU LOOK FOR IN A PROJECT LEADER?

Here is a list along with comments.

- *Problem-solving ability.* The most important attribute of a project leader is the ability to identify and solve problems. This capability in systems projects requires some technical knowledge, business knowledge, ability to work with people, and good analytical skills for evaluating problems. How do you detect this ability? Instead of asking someone about their achievements, ask what types of problems have they solved. After they mention some, then pose the following questions:

—What were the symptoms versus the problems?

—What alternatives did you consider in analyzing the problem?

—What decisions were made about the problem?

—What actions were taken from the decision?

—How did it turn out?

—If you had to do it over again, what would you do differently?

- *Learning curve.* Does the person have a learning curve where lessons learned and experience are reflected in better work as a project manager? How do you test this? Ask how the project the person is managing now differs from the one that he or she managed for the first time?
- *Seeing a project through.* A person can be a genius, but project leaders must have the willingness and stamina to see a project through to the end. Try to determine if the person changed between projects. Don't ask why the person switched. Instead ask, "Will you describe both projects—the one that you left and the one you went to?"
- *Being able to work with others.* In the past, this meant being able to work with technical staff. This is still important, but as important are abilities to work with other project leaders, line managers, and upper management. Here are some questions to test this ability:

—How did your recent projects interface to other projects? What issues arose between projects? How were these resolved?

—How do you foresee working with other project leaders? If you were assigned to help a junior project leader, what would you do?

—What is your approach for making presentations and communicating with line managers of business units? What kinds of contacts and what was the frequency of contact that occurred?

—What do you think are the most challenging aspects of managing highly productive technical staff?

—In what ways did you interface with upper management? How did you approach management to resolve issues?

- *Business experience.* In a modern organization, it is very important that a project leader have business experience. It would be helpful if the candidate has managed a line organization. Project leaders must be comfortable with business processes, personnel issues in line organizations, and managing business unit employees. This is at least as important as technical knowledge.
- *Technical experience.* This does not necessarily refer to programming or operations experience. You first have to ask how you would have the leader apply the knowledge. First, the leader would need to be able to understand and analyze problems. Second, he or she would participate in the project team by doing some of the work.

- *Project experience.* It is helpful to have a wide range of project experience, if possible. Here are some questions you might pose:
  —What were the longest and shortest projects on which you worked? What were the differences between these projects?
  —What was the most interesting project? What was the most boring project? Why?
  —What were the most challenging issues that you faced?
- *Administrative abilities.* This can come further down on the list. Included are knowledge of project management software and other tools, as well as the ability to do analysis, prepare budgets, and complete other similar work.

How Do You Find and Select Project Managers?

Here are some guidelines:

- Keep a list of potential candidates—both internal and external. Consider people in business units as well as those within information technology.
- Delay appointing a project manager until the project concept is defined. This will give you more information about the project so that you can pick the best candidate.
- Quiz the candidates based on the objectives, scope, and issues that surfaced in the development of the project concept.
- Create a table that compares candidates with areas of risk in the project. The rows are candidates and the columns are the areas of risk identified through issues in the project concept. In the table enter the numbers 1 through 5; the higher the score, the better you feel that the candidate can address the area of risk. You can also rate each area of risk on the same scale. If you now multiply the rating of the risk by the candidate's rating on that issue and then sum these, your total score should indicate how the candidate would be able to address the risks.

Once you have a list of candidates, then prepare a table such as the following. In the column labeled "Importance" put in a 1 through 5, where 1 is low and 5 is high. Rate each candidate on the same scale. Then multiply the rating for importance by the rating for the candidate to get a total score. In this example, candidate 1 is preferred based on this candidate's higher score. Had administrative ability been highly rated and problem solving lower rated, the choice might have been candidate 2.

Attributes	Importance	Candidate 1	Candidate 2
Problem solving	5	5	2
Learning curve	4	4	2
Seeing project through	4	4	3

Attributes	Importance	Candidate 1	Candidate 2
Work with others	4	4	5
Business experience	3	4	3
Technical experience	2	2	5
Project experience	3	3	3
Administrative abilities	1	2	5

WHAT ARE THE DUTIES OF A MODERN PROJECT LEADER?

You might expect to see a list of duties here. In systems projects, this does not make much sense because the duties vary depending on the phase of the project. Let's define duties according to the following phases:

- *Initial project kick-off.* This is the period that begins with the appointment of the project leader and ends when the plan is approved, the team is approved, and work has begun.
- *Preintegration phase.* This covers feasibility, analysis, design, software selection, and software development up to integration and testing. It also includes architecture and infrastructure work. It also includes work related to business processes.
- *Integration and implementation phase.* Covered here are integration, testing, business unit procedures, policies, training, conversion, and cut-over to the new system.
- *Postproject phase.* This includes closing out the project, gathering the lessons learned, and going on to the next project. However, note that lessons learned must be gathered along the way. If you want until only at the end, people will have moved on to other work.

Now you can divide the duties into categories:

- *Major duties.* These are the mainstream responsibilities on which project leaders spend most of their time.
- *Administrative duties.* These are the overhead functions that are significant in supporting the project.
- *Background duties.* These are duties that the project leader will be doing in the background. Examples are performing work in the project and ensuring that methods and tools are being used properly.

Why are these categories used? If you are going to be an effective project leader, then you must manage your time and be very aware of priorities among tasks in these categories.

You can now build a table such as the one in Figure 5.1. The rows are the phases and the columns are the categories of duties. Each of the table entries can now be discussed.

Phase\Category	Major	Administrative	Background
Kick-off	• Develop plan • Assemble team • Set methods and tools • Begin work • Resolve issues	• Set up project files • Establish electronic communications tools	• Collect information on other projects • Coordinate with other project leaders
Preintegration	• Resolve issues • Monitor actual work and quality • Perform marketing • Interact with business units and other systems managers • Resolve resource conflicts	• Update project status • Make presentations on **project** • Analyze budget vs. actual • Coordinate with business unit	• Perform work on project • Track status of related projects
Integration-implementation	• Resolve issues • Monitor actual work and quality • Perform marketing • Interact closely with business units • Resolve resource conflicts	• Coordinate with other projects • Update project status • Analyze budget vs. actual • Make presentations on project	• Perform work on project • Track status of related projects
Postproject	• Construct lessons learned • Support placing project members • Measure benefits of the project	• Close out project files • Prepare personnel evaluations • Close out presentation	• Ensure that team gets credit • Prepare cost-benefit analysis

Figure 5.1 Duties of a Project Leader by Phase

- **Kick-off**
  Major.
  —Develop the plan. Here you will develop the plan and gain management approval. It is assumed that the project concept has been developed. If it has not, then this would be an additional duty here. Included in developing the plan is the marketing of the plan as well.
  —Assemble the team. This includes finding the team members, negotiating availability, providing motivation, and building a team atmosphere.
  —Set methods and tools. There are two areas here. One is the methods and tools for managing the project. The other consists of those methods and tools that pertain to doing the real work in the project itself. Included here is the training of the staff in the methods and tools.
  —Begin work. Work on tasks begins. It is here that the project leader builds a pattern of how work will be reviewed and how it will be coordinated as a team effort.
  —Resolve issues. The scope covers identification, analysis, decision making, implementing actions from the decisions, and measuring the results of the actions and decisions.
  —Build communication with line managers and upper management for the project.
  —Assess and create ties with other related projects and work that impacts your project.
  Administrative.
  —Set up project files. This includes all project setup: paper files, electronic files, databases for issues, actions, and lessons learned, and so on.
  —Establish electronic communications tools. The project files must be established on the network. How electronic mail, groupware, the Internet, and other tools will be used are determined and implemented.
  Background.
  —Collect information on other projects. Your project does not exist in a vacuum. You must establish coordination, share project information, and collect information on other projects.
  —Coordinate with other project leaders. Get to know the other project leaders. Establish a working relationship with them.

- **Preintegration**
  Major.
  —Resolve issues. Same as above.
  —Monitor actual work and quality. You will have to monitor all work and determine tests of quality, completeness, and performance.
  —Perform marketing. This is a major duty because you must unceasingly work with upper management, business units, and other systems managers.

—Interact with business units and other systems managers. This is the detailed interaction on specific systems activities. In this phase, it tends to focus on the business process that is going to be supported by the new system.

—Resolve resource conflicts. Allocation of resources and follow-up are two major tasks here.

Administrative.

— Update project status.

—Make presentations on projects. This can include informal and formal presentations on issues, status, budget, and benefits.

—Compare budget to actual analysis.

—Coordinate with business units.

Background.

—Perform work on project. A project leader must do work in the project. Typically, this would be in the areas of systems analysis and design.

—Track status of related projects. You must determine the impact on your project of work and slippage in other projects.

- **Integration and implementation**
  Major.
  —Resolve issues. Issues here tend to be very important because you are getting closer to going live with the system.

  —Monitor actual work and quality. In this phase you are overseeing the work of the business unit staff as well as systems work in other projects.

  —Perform marketing. Marketing here includes change control and ensuring involvement in the project.

  —Interact closely with business units.

  —Resolve resource conflicts.

  Administrative.

  —Coordinate with other projects. Coordination today often means joint scheduling of work in this phase.

  —Update project status.

  —Compare budget to actual analysis.

  —Make presentations on project.

  Background.

  —Perform work on project.

  —Track status of related projects.

- **Postproject**
  Major.
  —Construct lessons learned. Although you are gathering lessons learned throughout the project, it is here that you have the time to put these together.

—Support placing project members. As the project leader, you owe it to the team members to help them migrate to other work.
—Measure the benefits of the project. You will develop statistics on the tangible benefits of the project.
Administrative.
—Close out project files.
—Prepare personnel evaluations.
Background.
—Ensure that team gets credit.
—Prepare cost-benefit analysis.

How Do the Duties Shift during the Life of the Project?

Note that there are many similarities between the preintegration and integration phases. The tone is quite different, however. In the preintegration phase, the emphasis is on internal development and work within the team. In the integration phase, the focus moves outside to the business unit and other systems areas. These differences point to the potential benefit of having separate or multiple project leaders because the preimplementation phase of a large project can be quite different and place different demands as compared to the implementation phase. This is a telltale sign of why some projects can appear successful during development and then fail miserably in implementation.

How can you use this information? If you have a project concept, then you can assess where the risk is. Is it in the earlier work or implementation? The answer to this question can be employed in selecting a project leader.

What Does a Project Leader Need to Know?

What do you need to know to be a project leader? It obviously depends on the nature of the project. Here are some examples:

- *Knowledge of the business process.* If the project is tied to a business process, then knowledge of how that process works, the exceptions, the policies, and the procedures can help.
- *Technical knowledge.* You should have some knowledge and at least limited experience with the systems and technology that are being employed in the project.
- *General project knowledge.* This is your background and knowledge of the projects in your organization. Included here is information on how the project management process works.

How Do You Become a Project Leader?

It is easy to be discouraged. Managers will tell you that they want experienced project leaders. Yet if you can't get experience, then what do you do? Here are some suggestions:

- In a project team, volunteer to take on issues. Gain problem-solving experience.
- Volunteer to take notes and perform project analysis. This will give you experience in different areas.
- Create several accurate versions of your resume for line and project assignments. As you will see, these are quite different. The line resume gives attention to the business process and number of people managed. The project resume focuses on the benefits, size, and hurdles overcome in the project. It is more difficult to get a project leader job with a line-manager-oriented resume.
- Come up with some project ideas and submit these to a friendly manager. The risk here is that the idea may be given to someone else. Therefore, pick some subject that people would not be attracted to. Here is a good idea for you: Gather and organize lessons learned from various projects. This is a task that many do not care to do because it appears to be too mundane.
- Get familiar with the project management process in your organization. Learn about other existing projects.
- Collect information on issues. This may prove valuable later in a project.

What Is the Role of a Superproject Leader?

In many situations there are several systems projects that are highly interrelated. How do you manage all of these if you have project leaders who are equals in each of the projects? Consider creating the role of the superproject leader. This is a person who oversees all of the projects in terms of dependencies and issues among the projects. The role does not include the detailed management of each project. That remains with the individual project leaders. Examples of superprojects are the implementation of an ERP (enterprise resource planning) system, process improvement, and e-business.

In more detail, here are some examples of duties:

- Perform analysis across projects in terms of resources and schedules.
- Identify and track issues that cross multiple projects.
- Conduct meetings among the individual project leaders to share information, resolve issues, and disseminate lessons learned.

What are the trade-offs here? Creating a superproject is probably a good idea if you have an application systems project linked closely to a technology infrastructure project. An example might be client-server applications with a wide area network deployment. On the positive side, you dedicate management resources to the projects as a group. This frees up management from involvement in the details. Issues among the projects tend to be resolved faster. Work quality may be better with more supervision. On the negative side, you incur the added cost. You also create problems when other projects, which are not in the group of projects, have to interface with a project in the group. Management can become too removed from the projects as well. Overall, is it a good idea? Experience indicates that a superproject leader concept should only be employed under exceptional circumstances when there are few dependencies with other projects not in the group.

In a superproject you divide up the project into large, but manageable subprojects. These subprojects can be based on responsibility, risk, or phase. In each subproject all of the responsibilities discussed before apply.

How Do the Responsibilities of a Manager of a Small Systems Project Differ from That of a Large Project?

Some might say that there are major structural differences. Experience indicates that much of the difference is a matter of degree. Issues are more frequent and important in larger projects. There are more people with whom you must coordinate and work. The technology can be more complex. There can be more interfaces.

Keep in mind one important factor—size is not as important as risk. You can have small projects with high risk and large projects that are routine. Another factor is time. If a project starts small, it can grow. If the work was begun without a project structure, then when this growth occurs, it becomes difficult to retrofit the work as a project. People on the team may feel that they are being penalized by having the bureaucracy of a project laid on top of them.

When should there be a project and project leader? From lessons learned, not every piece of work is or can be treated as a project. If you create too many projects, then you will spend much more time managing these projects and less attention will go toward minimizing risk. A guideline here is that work that involves multiple departments or key business processes should be treated as a project. If there is a short project within one department that impacts no other area, then it can be handled as work in the department with support from IT.

The responsibilities of the project leader for a small project then are the same as that of a large project. The goals are also the same: to minimize risk, to get the work done on time and within budget, and to ensure quality and performance. When you consider a large project, you want to consider the handling of multiple issues and interfaces at the same time. This will affect your selection of a project leader. However, it is not always true that junior project leaders are only suitable for small projects, because you can have large, routine projects that do not require extensive experience and expertise. An example might be deploying new versions of PC software across a large organization—a lot of work, but often with limited risk. The dominant factor is risk and the number, type, and severity of issues rather than size.

WHAT IS THE ROLE OF THE PROJECT COORDINATOR?

The project coordinator and the project office concept have been described in previous chapters. In either case, the modern role centers on the goal of ensuring consistency and quality of project management across all projects. Duties to support this include the following:

- Train and provide support to individual project leaders.
- Develop and maintain project templates.
- Monitor the project information and state of the projects.
- Support the identification and resolution of issues.
- Maintain and disseminate lessons learned across the organization.
- Coordinate the sharing of project information and resource deconflicting among projects.
- Perform project what-if analysis.
- Make presentations to management on specific issues, project status, and trends.
- Monitor the use of project management software tools and recommend upgrades whenever appropriate.

WHAT ARE THE ROLE AND SUITABILITY OF THE PROJECT OFFICE?

Moving up from the individual coordinator, you can expand the role to the project office. The project office performs the same duties as the project coordinator—it is a matter of scale. Some questions to answer are these:

- What is the number of projects that you anticipate over the next three to five years?
- Is the information systems unit going to provide project management support to other parts of the company?
- Are the projects significant in terms of business impact? Do they cross multiple departments and divisions?

The reason for the first question is obvious. The second question addresses whether the information systems organization is going to provide support outside of narrow information systems projects. The third question addresses the extent of communications and exposure that information systems has.

The project office entails more overhead and expense. Moreover, it runs a greater risk of being bureaucratic. With a project coordinator you can rotate the job function among different people. The project coordinator role is also useful in bringing new people into project management. This is less true with the project office. The danger here is to create a clerical, mechanical, reporting nightmare. The offset is that if you answer the previous questions positively and with a large number of projects, then you are more likely to benefit from a project office.

Here are some guidelines in managing a project office:

- Rotate the staff frequently in and out of the project office.
- Review all reports and requirements placed on the project office to ensure that the workload is necessary, meaningful, and has value.
- Segment the roles of the project office (see the project coordinator list) among different people so that you have accountability.
- Insist on measurement of the project office in terms of how long issues and so on remain in the project office.
- Allow project managers to approach a specific manager if they are running into problems with the project office.

CAN A PROJECT LEADER MANAGE MULTIPLE PROJECTS?

This is essential. There are not enough good project leaders to go around. Furthermore, more of the IT projects are intertwined and interdependent. This may lead you to think that managing multiple projects is good. However, follow some guidelines and lessons learned.

- Make sure that the projects do not conflict with each other for resources. Otherwise, you will find that the project manager is setting priorities on his or her own. Management has less visibility and might have a different opinion of what can be done.
- The projects should not involve widely different technologies. This spreads the manager too thin. For example, you might not want a manager to handle an intranet project and a legacy system project in which there is no interface between the two.
- The projects should involve the same business units. Projects that involve different business units face several risks. First, the manager must interface with more managers and so will be spread thin. Second, there are probably additional projects that interface with the same managers. This can create confusion with the business unit managers.

SOURCES OF FAILURE FOR A PROJECT LEADER

From years of observing systems projects, the following factor factors can cause the project leader to fail. Monitor yourself to ensure that you are not falling into any of these traps.

- *Dealing with issues as they arise.* This is a reactive mode and very dangerous in systems projects where there are many interdependencies. For example, if you make a decision on a client interface for a client-server system, it may have far-ranging impacts across all client-server applications. You must analyze issues in groups as opposed to individually.
- *Spending too much time in project administration.* This can cover many areas. You can spend too much time with management and not enough with the team. You can bury yourself in project management software. Your time is zero-sum—that is, when you spend it in administration, it detracts from the time in doing work and addressing issues.
- *Spending too much team contact time on status.* Project leaders who are not comfortable with the technology often retreat to hammering the team on project status. Consider the manager in the Dilbert cartoon strip.
- *Becoming obsessed with the schedule and percentage complete.* In these situations, status and progress take the front seat and quality and real performance are in the back. It is a problem because it just delays when problems will be addressed.
- *Micromanaging the project.* A technically oriented project manager may lean toward managing each detail of the project. One manager had all programmers submit code written during the week and reviewed it all on the weekend. It got programming results, but design and implementation issues were not addressed.
- *Failing to address outstanding issues.* Some managers work on the K of N reliability system. That is, if K out of the N issues are addressed, then they feel comfortable that they are on top of the situation. The problem really lies in the oldest outstanding and most severe issues. Not resolving these can have a major impact on the entire project.
- *Failing to get involved in the project work.* Even if you are not technically inclined, you can still perform some of the systems tasks related to analysis, design, testing, documentation, training, and so on. It is useful if the project leader works closely with the business unit staff. This shows that the manager is hands-on and gives the manager a better idea of what is happening in the business unit.
- *Tweaking the project plan or making too few changes.* On the one hand, you can drive people crazy by changing the project plan too frequently. On the other hand, you lose touch with reality if you don't change the plan to

fit reality. The recommendation here is to make fewer, major changes to the schedule so as to be flexible and provide stability.

- *Getting carried away with the technology.* The project leader who falls into this trap gets enamored with new technology and diverts the attention of the team toward the technology. After a few episodes of this behavior, team performance can be affected.

SUCCESS FOR A PROJECT LEADER

Several suggestions for success in the project leader role have been given already. Here are some additional success factors:

- *Continue to learn and grow technically with the project.* Many project leaders let their technical knowledge and skills become obsolete. Don't let this happen to you.
- *Be in frequent informal communications with line and upper management.* This includes not only addressing problems and opportunities, but also communicating interesting project stories and status.
- *Know what is going on in the project at all times.* This includes what people are working on, what issues are active, the status of the project, and upcoming milestones. Be prepared to answer almost any question on a moment's notice.
- *Work with team members individually.* Although team communications and work are very important, much of systems work is individual. Take care to keep in touch with each person as an individual. Go to your team members' offices and chat casually on a regular basis.
- *Use the team to identify issues.* Perform much of the analysis of issues yourself. Allowing team members to suggest issues that can lead to problems or opportunities gives them a larger role in the project. It adds to their commitment. On the other hand, be very careful about delegating issues to team members. They may have too narrow a perspective and lack organization and political knowledge.
- *Project management is political—don't forget it.* Across all projects, managing projects is and has always been political. Remember that projects typically mean change. Change can be threatening and disruptive. There can be resistance. Groups and departments may fight for resources. These situations are just a few examples of how politics enters into project management.

How do you cope with the politics? Experience reveals that this is one of the biggest challenges of project management. It is closely tied in with the ability to solve problems. In order to solve a problem, you must market and implement a solution. Politics also means getting along with people that may detest you or the project. Because of shifting roles during the term of a project, you cannot afford to alienate or get people angry. If you win an argument, you may later lose the war. Consider the following examples:

- *Line managers.* These people may have little interest in the project. They see it as disruptive. They may fear technology. Work with them and especially their people through the project to gain their confidence.
- *Upper management.* Frequent, informal contact where you convey status, issue information, and lessons learned will get their interest in the project and gain their support.
- *Systems managers.* Some may feel that you have a more interesting project. Others may resent the high priority that your project has because it robs them of their people.

How Do You Measure a Project Leader?

Here are some questions you might ask related to a project leader:

- How much time is being spent on the major activities versus administrative activities? Too much time on administrative matters will detract from work.
- What is the age of the most severe unresolved issue in the project? If it is a long time, then the project leader may not be addressing the issues.
- Is the project leader aware of the exact current state of the project?
- How often does the project leader communicate with upper management about the project on an informal basis?
- What is covered in project meetings? How long do the project meetings last?
- What is the turnover and mood of the project team? Do people on the team jump at the chance to work on other projects?
- What is the mixture of time spent on issues, communications, work in the project, and administration? The sequence here is probably the desirable ranking of priority.
- How does the project leader communicate with the team and departments?

EXAMPLES OF PROJECT LEADERS

Project Leaders in Manufacturing

One of the best and most capable project leaders encountered is in a high-technology manufacturing firm. He is one project leader among literally hundreds. Why does he stand out? He saw that there was a need to modernize the project management process across the company. He developed the concept and marketed it to management over the period of 18 months before it was accepted. He implemented the new process in about 35% of the company. It was very successful. It was built on as a result of many people accessing project information and updating their own schedules. The system allows everyone to see each schedule. The project leader also salvaged the project office and changed it from a bureaucratic, controlling group into a proactive supportive group. His success was rewarded with another project involving international communications. What were his strengths?

- He demonstrated initiative and creativity and developed many of the detailed ideas on his own.
- He stuck through the rough early days of the project when there was a great deal of resistance.
- He was successfully able to work with everyone from technicians to upper management.

Did he have enemies? Was the atmosphere political? The answer to both questions was yes. He ended up spending over half of his time marketing and handling issues. Toward the end of the project, he spent less than 10% of his time completing project tasks due to the need to overcome resistance.

GENERAL SYSTEMS PROJECT LEADER

This is a person who began work driving a taxicab in New York City. He was very street smart and savvy. He learned project management on his own while working in a bank. He then moved up to becoming a senior vice president. He also handled projects in insurance, manufacturing, and software development. He was very demanding, but very fair. He took no credit for anything and gave credit to the team. He built up tremendous loyalty among his staff. When he changed jobs, people clamored to follow him. His secret as a systems manager was to run all major and most minor work as projects. To him, everything became a project. He developed a sense of being able to detect risk.

E-BUSINESS LESSONS LEARNED

As was noted, e-business projects tend to be large. Moreover, they are really programs in that once you start e-business, you cannot drop the ball and say that the project is finished. Ongoing work is required to keep the firm competitive. For these reasons, you should consider having two project leaders over the implementation overall. One should be business focused. The other can be IT focused. Why do you need two project leaders? One reason is that e-business is so important that you cannot afford to not have backup. Second, you need a very broad range of skills and knowledge. It is not likely in a big project that one person can handle it. Third, since it is a program, you will need to have someone continue the management of e-business work after the initial implementation is completed.

GUIDELINES

- Give credit to the team members. Never take credit yourself. You will get credit as management recognizes the project. Management first tends to recognize the work in the project. Then managers consider the project itself. If you carry this off right, then you will be praised by the team members.
- Build a reputation for measured action—that is, perform a great deal of analysis on systems issues. Then when the decisions are made on the issues, put the supporting actions into place quickly and with determination. If you merely act without sufficient analysis, you will acquire a reputation as a shoot-from-the-hip gunslinger. If you analyze and decide, but fail to act, then you will be viewed as indecisive.
- Establish ties to multiple managers in upper management and business units. Don't rely on just one person. That person may change jobs during the project. Also, you will to sell ideas to multiple managers. If you sell to one person, you are then relying on that one person to carry the day.
- Trade off youthful enthusiasm versus experience in selecting a project leader. Both have advantages. If you give this careful thought, it will help you to understand the project better.
- If you are project leader, be alert to the fact that you can be overused and abused. In some industries such as consulting, this can result in burnout. This is particularly true for project leaders who gain a reputation for being able to turn around a project. They continue to be assigned failing projects. What do you do if this happens to you? You must have the guts to jump ship. Keep your resume handy and look for new opportunities.
- Linked to the preceding point is that when you are hired by another firm as a project leader, people's perspectives of you will be fixed quickly. Your first actions as project leader will be watched closely. Be careful.

WHAT TO DO NEXT

1. In your organization, analyze how people become project leaders. Are the same people always selected? How do these people keep up their skills? Are there some common skills among the project leaders? For example, are they all good at problem solving?

2. Suppose that you want to be a project leader. Which projects do you think you can manage? It is important to analyze why you feel that certain projects are more suitable than others. This will tell you a lot about yourself.

3. For a project that you are involved with, what ties exist between the project leader and the business unit managers? How do you assess the communications with business units on the project? What issues can you trace to communications problems with the business units?

SUMMARY

The best project leaders often got there by working successfully on a series of projects. They turned some around and successfully completed others. They may have had good luck, political support, and other things in their favor, but in the end it is often hard work and perseverance that pulled them through. When you become a project leader, you must constantly measure yourself along the lines suggested in this chapter. Be your own worst critic.

Chapter 6

Building the Project Team

INTRODUCTION

A team in a systems and technology project is unlike standard teams in several respects. In a standard, traditional project, the team members tend to stay on the team for the duration of the project. The scope of the project tends to be more stable and fixed. People on these projects do not have to perform nonproject tasks in many cases. Tasks are interrelated.

Managing teams in systems projects is more challenging due to the following factors:

- The team members are more in demand by other projects, as well as by companies who want to hire team members away.
- Many team members only handle certain parts of the work and then they are gone.
- As new phases of the project begin, the requirements on the team change.
- Many teams members have to do their normal business or IT work in maintenance, operations, and enhancement.
- There are more opportunities to work on joint tasks as well as the more parallel effort.

Due to these factors, it is more difficult to manage the teams in such a fluid situation. Project leaders spend considerable time finding new members, releasing team members, and building teamwork—more than they do for standard projects.

Even with the differences, there is some common ground. Today, organizations and projects move to *empower* team members to define their work and how they are going to go about it. Team members are also held *more accountable* than in systems projects of 20 years ago. Motivation of team members and the building of team spirit are important for the core members of the team—the people who are going to be on the project for a long time.

Earlier we recommended that you have small project teams. Even in large projects, consider establishing a small core team of no more than three to four people. These are individuals who are generalists and problem solvers and who can deal with issues in different phases of work across the project. The lessons learned for managing the core team resemble those learned from managing a traditional project team. Carry out much of your planning and problem solving with this team. Examples of core team members, in addition to the project leader, include the following:

- A systems analyst or designer who will support initial requirements through testing and implementation.
- Key programmers who will develop the software for the system and interface it to the existing internal software.
- Business unit staff. There are several roles here. One is a coordinator who acts to control changes, handle coordination, and arrange for resources from the business unit. This is a higher-level person. Another person will address lower-level tasks associated with requirements and the business process as well as training, testing, and documentation.
- A network or operations person. This person is responsible for overseeing the hardware and network as well as the nonapplication software.

These roles are most often filled by internal employees. However, a contractor or consultant may be most suited to some of these roles.

There are pros and cons for smaller and larger core teams. Smaller teams reflect the problem of getting good people. A small group is easier to manage and coordinate. There is less chance of having people around who are underutilized. However, if a member of a small core team leaves, there is a major gap to fill. If people in a small team cannot get along, then there can be more problems than in a larger team where you can attempt to isolate the problem. In some organizations, a larger team means more power for the project. This is less true today, but it still holds.

Moving to the project team overall (called the general project team in what follows), you will find roles for many more people. They include bit parts, walk-on parts, and part-time parts in your project cast. Slightly different but consistent rules apply here. Examples of roles include the following:

- Network installation and testing
- Hardware and system software installation and testing
- Additional business unit staff to support testing, documentation, training, and so on

- Other programmers who support the systems with which the new system is to interface
- Operations staff to support testing and implement production
- Contract programmers
- Consultants
- Business staff who are energetic and desire change
- Business staff who have extensive experience and knowledge of business rules, shadow systems, and exceptions

APPROACH

WHAT ARE THE RESPONSIBILITIES OF TEAM MEMBERS?

Each project member has a responsibility to define the detailed tasks that will be needed to perform his or her work. The individual must define what methods and tools must be employed to do this work. Flexibility is desirable so that these will conform to the methods and tools endorsed by the organization. Each team member can establish his or her detailed tasks in the project plan on a network. Team members should be able to update their respective parts of the plan. In addition to supplying status information on the work, they can participate in reviewing other work in the project. These remarks apply to both the core and general project team.

The core team members have a wider range of responsibilities in the project. They identify issues as well as potential solutions to the project leader. They must take responsibility for investigating issues and proposing solutions. They act as an early warning system for any potential problems that arise outside of the project team. As people with general skills, they must be able to stand in for each other. If multiple business units are to be dealt with, then some members of the core team might be very involved in working with them. Core team members must review the project plan on their own to determine potential problems. This is valuable because they have different project and technology experience than does the project leader. In some cases, core team members oversee the work of the part-time team members.

WHAT SKILLS DO YOU REQUIRE FOR THE TEAM—BY PHASE?

Project Start-Up

Ask yourself here and at each step, "Where is the risk?" At the early stages of a project, the risk is in not getting the requirements and business transactions defined properly. The risk is also in underestimating what will be required in terms of infrastructure (such as hardware, network, or software). A recommendation is to ensure that the project is covered in terms of systems analysis. This will be helpful for developing requirements, sizing technology, and identifying what is required.

Preimplementation

Risk here can lie in development, interfaces, and guarding against changes in requirements. If the analysis performed at the start of the project is not complete, then additional business rules and exception transactions will now surface. These can have a severe impact on the schedule because they were unplanned. The core team must not only perform the work in this phase, but also monitor and ensure that the scope of the project does not begin to expand. Required here are technical skills related to analysis, design, and programming. In addition, the team must address issues that arise related to the business unit and respond quickly to any technical questions in order to avoid delays.

Implementation

During implementation, many of the technical skills continue to be required from preimplementation for support of integration and testing. Many new skills are needed. These skills relate to procedures for the business process and system, testing, documentation, training materials, operations, and training. Once errors or problems surface, the expanded team must move to resolve these. If they cannot be handled through programming in the system, then there may be pressure to devise a strategy for working around the system in order to cope with the exception.

WHEN SHOULD YOU FORM THE TEAM?

For the core team, identify people during the development of the project concept. These are the individuals who will be brought on board first. For the general project team, wait. If you form the team too early, you can run into problems. There may not be work for them to do. Morale will then suffer. If you bring on people before you have coalesced the core team, you risk disturbing what you are trying to accomplish with the core team. Err on the side of assigning people slightly late and then make up the small difference in the schedule. Call this just-in-time staffing. Note that in a traditional project you want to establish the team early. This has several drawbacks in modern technology-related projects. First, you really do not know who are the individuals most suited to the project. Second, if you get people too early, there may not be enough for them to do.

Note that the general project team will never be completely formed because it is always in transition. Tasks for the part-time people tend to last at most several months. As the work progresses, there will be overlap as people enter and leave the project team. Identify these part-time members according to the work that is to be done rather than trying to obtain specific individuals. Traditional project management often stressed getting people early. However, at the start of an IT project you may not have information on what you need and on who the most suitable people are. Moreover, if you get people too early, they may not be productive on the project.

HOW DO YOU GET TEAM MEMBERS?

For the core team, you can rely on your experience and what you know about the project. You will have to negotiate with the managers in systems and the business units. Prepare a brief informal presentation about the project, which will answer the following questions:

- What is the purpose and scope of the project?
- What is the role and importance of core team members?
- Why is the project important to the organization?
- If the project is not successful, what can happen in a business sense?
- If the project is successful, what are the benefits to core team members?
- How will their work be managed?
- What flexibility is there for them to return to their organization?
- Identify and discuss the other work that the team members still have to perform and suggest how resource allocation will be done.

This last question is important because you appeal to the team members' self-interest. Two things a project manager can provide are support and exposure— that is, you can give them opportunities to present the results of their analysis and work to management. A good project leader gives the core team and some of the part-time members this opportunity for recognition.

Will you get all of your first choices for the project team? Most likely you will not. A method here is to approach a manager and indicate your requirements, schedule, and minimal level of capabilities needed by your team members. Let the manager suggest several names. For the part-time team members you will have less political clout. Recognize this and live with it.

SHOULD YOU USE CONSULTANTS?

Information technology has long been an employer of consultants and contract help. Unfortunately, however, organizations often decide on consultants on an ad hoc, one-event-at-a-time basis. Their use for individual tasks may make sense for control, but it also raises costs and can yield inconsistent results. This approach has a number of drawbacks:

- Use of consultants is uneven within information systems, causing the internal staff to become confused as to the role of consultants in projects.
- There are no ground rules for evaluating, selecting, and terminating consultants.
- Without an overall approach, the use of consultants, the choice of consultants, and other related factors are almost entirely based on political considerations.

Why employ consultants? Here are some of the political and technical reasons:

- You lack specific technical knowledge. You will use the consultant to provide this knowledge and train your people so that they can be self-sufficient.
- If you use consultants, managers may believe the consultants more than they believe internal staff members. This is regrettable, but it does occur. Sometimes, you find that credibility is directly related to the hourly rate: the more you pay per hour, the more credible the consultant is to some managers.
- You require a consultant who knows the business and industry. This is increasingly common due to reengineering, e-business, and client-server systems that fit with the business process.
- You are overcommitted to projects. Consultants can augment your staff. An example might be the use of consultants to do software maintenance.
- Due to various circumstances, you find that you are unable to attract and retain qualified people. Consultants provide a means to continue work.
- You are interested in saving money. Consultants are sometimes less expensive than employees when you factor in overhead and benefits. You also avoid a long-term commitment.

To establish an orderly approach for consulting, first develop a set of objectives for your systems. After matching your staff to these objectives, you may find that you want to consider consultants. What are some potential objectives?

- To provide low-cost basic services related to information systems. This approach will likely result in few consultants because you will be operating in a maintenance mode. There will be little development.
- To establish major new systems and technology capabilities quickly. This objective will require more consulting because your current staff members cannot handle their current work along with this new work.
- To ensure flexibility in staffing. Here you might be ready for business mergers or acquisitions and wish to retain flexibility to add people later. Therefore, you will use consultants to provide that flexibility.
- To concentrate your internal resources on only core systems activities and business processes. The overflow, so to speak, can go to consultants.
- To establish an approach for allocating consulting resources during the project.

When considering consultants, you must examine constraints or barriers. One is the company policy related to the use of consultants. What are specific rules and restrictions? Another is your budget—not just the total budget but how much you have set aside for outside services. A third constraint is the availability of consultants with the specific skills that your projects require. The project itself also plays a role. If the project involves company-sensitive information or will yield a system that provides competitive advantage, then even with signed legal agreements, you may be at risk of having the information and expertise move to a competitor.

Develop an overall strategy in line with the objectives you set. Here are some examples of strategies for consultant use:

- Employ consultants to maintain current systems so that the internal staff can develop new systems. This strategy sounds nice, but it may be difficult to pull off due to the system-specific knowledge associated with legacy systems.
- Apply the reverse of the first strategy—use consultants on new development and then migrate the current staff to the new systems later. This has the risk of lowering morale of the internal staff because the consultants will get all of the "interesting work."
- Augment project teams with only selected consultants where there are short-term needs. This can be dangerous because it can be open ended.
- Determine areas where you want to outsource certain systems activities on an ongoing basis. Examples might be PC training and hardware support.

How Should You Hire and Direct Consultants?

The biggest mistake people often make is to approach this backward—that is, they go out and find a consultant and then they define the assignment and work. A more organized method is to follow these steps:

- *Step 1:* Define your objectives and strategies with respect to consultants.
- *Step 2:* Identify the target functions, activities, or projects that consultants are suited to within the strategy of step 1.
- *Step 3:* Develop requirements for the consultants in terms of a statement of work.
- *Step 4:* Define interview and evaluation criteria and identify members of the project team that will conduct the evaluation.
- *Step 5:* Identify and contact potential consultants.
- *Step 6*: Conduct the evaluation and selection.
- *Step 7:* Negotiate the terms of the agreement.

The statement of work describes the project, the role of the consultant, the specific tasks that the consultant is to perform, the end products expected from the consultant's effort, the schedule of the work, and the methods and tools that will be employed. This does not have to be formal. It can be developed as a series of lists. The statement of work provides focus and can help potential consultants explain their qualifications in terms of the work.

Gather the team members who will conduct the technical interviews in a meeting to discuss how they will perform the interviews and the questions they will ask. Particularly important is the validation of experience and capabilities. Encourage questions such as, "How would you address . . . ?" By providing situations in programming, analysis, and other areas, you can get the person to open up and provide you with more information. Team members can develop a list of questions in advance.

To find consultants, you can request referrals from friends and contacts. You can also search the Web and Internet for potential firms. Another source is magazine articles. What are the types of firms and their advantages and disadvantages? If you hire a large firm, then you have less control over the individuals who will be assigned to the project. You do get the benefit of gaining access to a wider pool of people. However, larger firms may carry a greater overhead and burden rate. This can put them out of your budget. Some firms and individuals are probably preferred for limited-scope projects. You will obtain a lower cost and you can be assured that the people you interview will be the ones who actually do the work. However, if they leave the project, you may be at risk and may have to search for someone all over again. Because the new people lack backup, you probably will have to assign your staff to work with them closely.

In evaluating and selecting consultants, select at least two or three. This will give you backup in case a problem arises with the leading contender. In some projects, you may want to employ several firms. For example, if you were implementing local and wide area networks in several locations, you might pick a local firm in each city. If you were implementing many PCs in one location, you might also divide the work among several firms. This involves more coordination and management time, but often can lower your cost.

On a specific development project, there is some benefit in employing several individuals from different companies. You are provided with different points of view. You also may be provided with better service because the individuals are competing for potential additional work later.

TEAM DYNAMICS

At the start of a project, the project leader must build team spirit—people on the team work together on their own without direction from the project leader. You must set the stage by discussing the following aspects of the project with each person individually:

- The purpose of the project and why the person's role is important
- The scope of the project and how the individual will fit within it
- Why the person is on the team
- What you expect of the individual
- What he or she will get out of being on the project
- How you will support the individual and provide for training and tools
- How you want team members to work together; give examples
- The individual's role in project management with respect to status and issues
- What you want to do for the individual

Ask if the person has any concerns. Be able to respond to the following questions:

- If the project is changed or cancelled, what will happen to us?
- Will the right tools be available for development?
- What are the major risks in the project?
- What is the current status of the project?

In terms of benefits, each team member will have the opportunity to do the following:

- Implement new technologies and gain new skills
- Add a successful project to your resume
- Have the opportunity to work with new people and gain contacts
- Gain the help of the project leader in advancing his or her career after the project has been completed

After this, you can hold a team meeting and go over the following:

- Review the purpose, scope, and project concept with the team.
- Reiterate roles of each team member and what is expected of each one.
- Identify how you expect the team members to draw upon each other.
- Identify how project meetings will be conducted and held.
- Review the benefits of the project to each member of the team from each individual's perspective.

SHARING TEAM MEMBERS BETWEEN PROJECTS

In a modern project, you will be increasingly sharing members of the project team with other projects and the line organization. What are your goals? You want the person available to your project for the specific tasks for which he or she has signed up. To ask for additional time and effort is probably unrealistic if the person is in demand. Here are some guidelines on resource sharing:

- Negotiate availability with the line manager or other project leaders at the start.

- Indicate which tasks that you will need the team member to be involved in and specify the general schedule.
- Establish a process for notifying the manager in advance as to when the person's services will be required.
- Indicate that you will monitor the work and release the team member as soon as possible and that you will provide informal status information.

COMMON TEAM PROBLEMS

In a technology project, it is possible for the team to become locked into a specific method, tool, or approach. This is not unusual because the team is living so close to the work under pressure. Here the project leader can play a unique role. The project leader is often the only one on the team who can take a wider perspective and sit back and ask some basic questions such as "Is there a better way to organize the work?" "Should assignments be changed?" or "Should the project approach be altered in a major way?"

You can assume that the people on the team will talk about the project with individuals in their line organizations. This is natural and you cannot control it. However, this tendency opens the potential for damage. If the wrong impression of the project leaks out, then you, as the project leader, will have to spend hours turning the situation around. At the start of the project, address communications outside of the team. Indicate that if anyone has a problem or issue, you want to know about it first. Point out the potential damage to the project and team if disinformation is disseminated. Follow this up during the course of the project by visiting each team member and asking what is on his or her mind. What concerns do the team members have? Do this in every meeting.

You may be forced through politics or circumstance to take someone onto the team that you feel is not qualified or has some problem. You can attempt to fight it, but doing so can be counterproductive. The person will find out how you feel and that will cast a pall on the project. If you are thinking that this situation will occur, then define some initial small tasks that you can monitor. Let those team members who you are concerned about perform these tasks until they prove themselves or fail. Then you have grounds to replace them. This method will also minimize the impact on the project.

Team members may not get along with each other. Over half of the projects that the authors have been involved with have had this problem. Expect it, because many technical people have their own opinions and methods of doing things. If conflicts arise, attack specific questions and disagreements as issues. If the underlying problem is severe, then consider gathering the people involved in a room for a meeting. Indicate that you have observed communication problems, that this is to be expected, and that you expect them to behave like adults and discuss issues openly. Don't insist that they get along.

Technology projects burn people out. These projects have tight deadlines and other pressures. You cannot run a project as a nonstop "death march." This can kill the project. Instead, plan when the team will have to perform on a superhuman scale. Plan also for periods of slack. Much of the pressure in projects is generated by the project leader or manager. If you are asked to speed up a project, ask what is behind the request. What are the business reasons and what are the benefits from the acceleration? Remind a manager that this can only work for a short time. The pressure will have to ease up. If you fail to follow these recommendations, then you will lose control of the project, the people will become burned out, and the overall project may fail or slip even more.

Much project work is routine and seemingly boring. For example, many programmers don't like to do testing and integration. They prefer development. Recognize that this can happen. Plan ahead to monitor the work during this period more closely. You definitely want to show interest in their work at this time.

In many small team systems projects, you are dependent on one or two people for many critical tasks. Sometimes, in response, a project leader will ride these people and drive them nuts because the leader feels that these team members are so critical. A more laid-back approach may yield better results. Indicate to them that they are critical. Ask if you can provide any help or extra assistance. If they decline, they should still feel better because you showed that you cared about their welfare.

Problems can arise when new technology is adopted in a project. The team members may not know what is expected of them with respect to the technology. They may feel that there is even more pressure because they have just acquired the added task of learning the technology, although the overall schedule remained the same. What do you do? Before the technology is deployed, indicate the impact on the schedule. Explain what is expected of team members in reference to the technology. Discuss how you will provide for training and support. Have several people learn the technology and teach others. Do not immerse everyone in it. That will destroy productivity. Establish someone on the team as an expert or at least someone that the team members can go to with questions.

Systems projects have many interdependent tasks between people. This can lead to problems if someone waits for another person to finish a task before he or she can work. Head this up during project planning by defining each hand-off and explaining how the individuals can continue to work while waiting for the results of the predecessor task.

Management may seek to change the direction of the project and even the composition of the team in the middle of the project. How do you react? You should never have gotten into this position in the first place. When the project is approved, point out the potential impact of change then. Reinforce it through the early phases of the project prior to implementation. Once you reach implementation, then the momentum will likely keep the project going. Next, have some early milestones and implementation that shows success. Management will be less willing to make changes. Third, ensure that there are sufficient milestones along the way that can help buffer the project from change. Also, build support for the project through the business units and indicate to them that there is always the threat of change and stopping the project.

MANAGING TEAMS

The project leader is responsible for building and maintaining a team culture on the project. Lessons learned show that in systems and technology projects this is very important. The culture determines how people will share information, bring up issues, and resolve issues among themselves. Team culture also impacts the extent of cooperation. These are critical success factors in the project, especially because almost all modern projects involve substantial integration within systems and with business units.

What is the project culture? Here are some determining factors:

- *How team members work with the project leader.* This includes how they bring up issues, how they report status, and how they share information with the project leader.
- *How team members work together.* In addition to performing work, this encompasses how team members work together on issues with and without the project leader.
- *How team members perform their own individual work.* Included here are the sharing of lessons learned, the organization and approach to work, documentation, and how one's work is passed onto others in the team or external to the team.
- *How the team deals with crises.* This seems to be unpredictable, but in modern systems organizations you will find the same crises occurring again and again.

How do you build and sustain the culture? Here are some guidelines:

- Establish some basic rules at the start. Define each person's role with respect to the team as a whole. Give examples of interaction.
- Reinforce the need for cooperation by singling out examples of cooperation throughout the project. The problem is that many project leaders give praise to individual achievement, which reinforces the importance of individual work over teamwork. Alternatively, some project leaders give general praise to the team. This is so vague that it appears meaningless and has little impact.

- Simulate some crises to the team in a meeting. An example might be to propose a major change in requirements. Have the team attempt to deal with this hypothetical crisis. This will build teamwork and will also get the team ready for a crisis when it occurs.
- The project leader should meet with not only the individuals and the team as a whole but also with groups of several people on specific areas of the project. The project leader must sustain the team in subgroups.

How Do You Keep Team Members?

Team members can lose interest in your project. They may have competing demands from other projects. Assuming that you do not want to keep them through financial means, here are some other suggestions. Involve team members in identifying and resolving technical issues. The greater the involvement up to a point, the more they become committed to the project. If you overinvolve them, then they can get turned off by the politics.

A basic lesson learned in project management is that many team members detest the overhead and bureaucracy of project management methods. They hate the endless status meetings. They don't like being hassled about issues. Consciously attempt to minimize the hassle factor. In meetings, focus only on issues and work. Reserve status and updates of plans for short one-on-one meetings. Minimize the amount of paperwork that you require of team members. The project leader keeps the administrative side of the project away from the team. Complaining about this to the team can impact morale.

People also resent being kept in the dark. Many technical people don't like surprises. For example, let's suppose that you expect management to request that the project be speeded up. You can wait until management makes the announcement and then pass it on, but people can get real angry about this. On the other hand, you can alert team members to the possibility. Then you can ask them to plan their work for this contingency and see what they can come up with on their own. Later, if management makes the decision to speed up the work you have reduced the effect.

A commonsense suggestion is to show interest in the team's work. Don't just ask for status. Instead, have members describe it in more detail to you while you visit their offices. Ask them if they need any help or resources.

How Do You Discharge and Replace Team Members?

What if you have removed someone from the project? You can assist people who have done their work well by helping them in transitioning back to their organizations or to other projects. For people who presented problems, then work to find them their next homes or projects first. If you confront them and indicate that they will be leaving, they may do damage to the team and project during the time between your announcement and their actual departure. When someone does leave, announce this in a project meeting with everyone present. Indicate the person's achievements, new position, and how his or her role will be filled.

You have several choices with respect to replacement. You can move to replace someone immediately. This is the common method, but it has several shortcomings. First, it assumes that the new person has exactly the same skills and similar technical knowledge of the person who departed. This is almost never possible in systems and technology. Second, you are missing an opportunity to reformulate the project. Third, you are not giving the new person an opportunity to put his or her stamp and organization on the work. Many programmers, for example, resent being thrust into some situation where they must inherit someone else's work.

A better method of replacement is to leave the position open and start recruiting. When you have narrowed the field down to two or three people, then go back, look at the plan, and determine how you may change the plan and schedule to take best advantage of each person's skills. In systems projects a major personnel selection criterion, in addition to skills and knowledge, is how an applicant fits within the project. Each finalist can be interviewed by key members of the team. After the interviews, answer the following questions:

- What does the person add to the project that is not there now?
- In what areas are there gaps or holes if this person is brought on board the project?
- How would this person get along and interact with the project team and the culture in the project?

EXAMPLE: ASTRO BANK

The management of the project teams for both the legacy and new systems started badly. Separate project teams were established for each. No roles, responsibilities, or rules were established. The project leader assumed that everyone had previous project experience and knew what to do.

The projects began to deteriorate. The new system project team had never worked together before. Many were junior people who lacked project management experience. The work was not getting organized. The team just started in and began on tasks. The legacy system team was the reverse. Team members had operated in a bureaucratic project management mode for years. However, the projects were merely maintenance, operations support, and minor enhancements. Their approach was ill suited to development and major interface work. They treated the new work like the old. Their project meetings were almost social occasions.

The project leader faced change or failure. The first step was to impose project management similar to that described earlier on the new team. Members of this team were the least resistant and most accepting of organization. Once this process was in place, it was time to turn to the legacy system team members. There were two alternatives. The project leader could attempt to invade the existing culture in their organization. This was seen to be very risky and would run into management as well as staff resistance and dissension. The project could afford neither. The alternative was to involve the legacy system team members as individuals within the new system project team. This was the first step. Over time, the project leader then carved out a subproject for the legacy system team and ran the group under the same rules as the new systems project team. Meetings were held in the offices of the new systems development area—away from the legacy system staff.

E-BUSINESS LESSONS LEARNED

The guidelines and approach in this chapter really apply to e-business projects. The projects greatly benefit from having many users involved at different times. Also, as work progresses, different skills and knowledge are required. Critical business users are important to the success of the project because of their knowledge of business rules. For IT staff it also applies because most members of the IT unit have other systems duties and responsibilities. While it is nice to consider assigning people to such an important project full time, it is impossible to ignore or drop work that keeps the core systems and technology operational.

A critical success factor in e-business is resource allocation. A regular method is necessary to dynamically allocate IT, business staff, and consultants. Consider doing this on a weekly basis in a meeting with business and IT managers.

GUIDELINES

- In most projects you seek to have as many business users involved as you can. Why? Because the people involved will be supportive of change later.
- Have team members fill in detailed tasks under your summary tasks. Involve the team in hands-on updating of their schedules. These actions will increase their level of awareness and commitment to the project.
- Make all issues and project plans available to project team members at all times. If you attempt to hide or make it difficult to get information, it will only create doubt and mistrust.
- Consider dividing the project team in both the preimplementation and implementation phases into subteams. The use of subteams does not necessarily add bureaucracy to the project, if done informally. Each subteam can focus on a major milestone. This makes it easier to manage the subteam as opposed to individuals. Another benefit is that you can organize and monitor discussions and interfaces between the subteams.

- Keep in touch with the line managers who supplied team members to you. Remember that the only other communications are word of mouth, second-hand information, and what the team member provides to them in the way of feedback. This can be disorganized and lead to the wrong impression. Keep in regular touch with all line managers by providing them with informal status and issue updates. This will keep them interested and provide a context for understanding other information that they receive.
- If you have a choice between a highly experienced person and a junior person for the project, which do you choose? Conventional wisdom might be to select the experienced person. This is correct for highly technical work. However, many tasks in a systems project do not require this skill level. Seriously consider a junior person who is willing to learn.
- Institute an apprenticeship method—that is, pair junior people up with senior people on some tasks. The senior person may resist on the grounds of productivity. However, make it clear at the start of the project that this is part of cooperation. Indicate what your expectations are for the apprenticeship and what each role is. The junior person must perform specific tasks and cannot just be a part of a two-person subteam.
- Team communications can be taken to extremes. If some of the people have worked together before on different projects, they may socialize too much. Therefore, it is important to keep project meetings focused on technical and business issues. Otherwise, the conversation can drift off into discussions of the past.
- When the project adopts a new method, tool, or technology, anticipate resistance and even some hostility. Some programmers and analysts resent having to learn something new when they feel that what they have been doing for years still works. This resistance is behind the failure of many "structured" techniques in the past.
- If a team member is faced with many tasks, it is natural for the person to work on the easy tasks to show early progress. This attitude is reinforced if the project leader keeps emphasizing progress. For most systems projects, it is better for the person to work on the more difficult tasks that carry risk and uncertainty.
- A project leader works side by side with the project team members. This is more than just sitting in on meetings and providing feedback. By participating in the work of the project, the project leader builds respect from the team.
- If the roles and work processes within the team are not defined and enforced, then the team members will not know what to do. They will start inventing new methods. These can then become habits that are difficult to break.
- At the start of specific tasks, have the team members discuss how they would assess quality and perform testing. Delaying this discussion until the work is done can negatively impact the schedule. The person who did the work may become defensive then as well. At the start of the task,

there is not the sense of ownership of the work, so the discussion can be more open.

- Assign each detailed task in the project plan to one person. This will ensure accountability.
- Encourage and collect lessons learned during the project. Don't wait until the end. Ask, "What have we learned in the past week?" Ask this question regularly. The project leader can maintain this information.

WHAT TO DO NEXT

1. In your recent projects, were the roles and responsibilities of team members clearly identified? If not, what was the impact of this lack of definition?
2. Build the following table for several projects that you have observed or in which you participated. This will help you to determine some successful and less than successful methods.

Project	Team approach	Results and comments

3. Conduct the following survey of a project. In this list a 1 is low and 5 is very high.

What was the turnover of staff in the team?	1 2 3 4 5
What percentage of meeting time was spent in getting status versus discussing issues?	1 2 3 4 5
What level of orientation was given to new members of the team?	1 2 3 4 5
What was the level of participation of the team members in defining their own work and tasks?	1 2 3 4 5
What was the level of participation of the team in project management activities?	1 2 3 4 5
What was the level of involvement of team members in presentations before management?	1 2 3 4 5
What was the extent of information provided by the project leader to the team on management communications?	1 2 3 4 5
What was the level of cooperation between team members?	1 2 3 4 5

SUMMARY

Managing systems and technology project teams is often more complex than that of standard projects. There are more interfaces and the need for a greater degree of cooperation. The resources on the project team are often required by other projects. New methods and tools as well as new technology products are constantly emerging. The best approach in team management is a mix of formal and informal methods. Formal methods include defining roles and responsibilities, establishing rules in the project, monitoring cooperation, and ensuring participation. Informal methods include working with the team members individually and in small groups as well as in sharing information.

Chapter 7

Developing the Project Plan

INTRODUCTION

In traditional project management, you developed the project plan when you defined the project concept. However, this is too early and locks in the project too soon. Using the method here, the project concept has been approved by management and is ready for support with the detail of the project plan. Many think of a project plan as a dry list of tasks with resources and schedules. For systems projects it is much more. It is a political document that can assist you in communicating with business units and in gaining support for the project. Ask yourself how a business manager can understand a technology project with no technical knowledge—through the project leader, the project concept, and the project plan.

Another misconception is that the plan consists of only the schedule and costs. It is much more. What and how information will be managed are part of the plan and approach. These are significant because in systems projects you desire to have greater standardization across projects, as discussed earlier.

The goals in developing the plan include the following:

- Determine how the information relating to the project will be managed.
- Define the project plan itself.
- Develop budget estimates.
- Link the plan with those of other projects and work.
- Assess risk in the plan.
- Market and sell the plan to management, the project team, and the business units.

Notice that the scope includes marketing and sales. You can develop the most wonderful plan, but if you lack backing and support, you and the plan lose.

APPROACH

WHAT INFORMATION MUST BE MANAGED?

Beyond the plan and schedule—and the work—you must define the method for coordinating issues, lessons learned, and action items to resolve the issues. All of this must be integrated with the project plan. How can this information be related?

- Once an issue has been identified it must be related to the specific tasks in the schedule to which it applies. Otherwise, how do you determine the impact of not resolving an issue on the schedule? In turn, you must be able to move from the tasks to the list of issues.
- After a decision has been reached on one or more issues, actions must be taken to resolve the issues and implement the decisions. These are to be tracked. The actions may involve changes to the plan. It is useful to indicate in the revised plan why and when these changes occurred.
- If you have built up a body of lessons learned, then the challenge is to employ them. Lessons learned must be related to the tasks in the same way issues were.

Implementing these procedures in the context of project templates creates greater standardization and supports more effective project analysis.

WHAT METHODS AND TOOLS WILL BE EMPLOYED FOR PROJECT MANAGEMENT?

Two sets of methods and tools are used in a systems project: one for project management and one for the work in the project. Areas for project management are as follows:

- Team communications
- Issue tracking and information
- Lessons learned information
- Project plan information
- Project documents
- Integrated project information (all of the above in one place)

A tool is the software package that you will employ for an area. A method consists of the procedures for carrying out the coordination and task. It is best if the methods and tools are in place at the start of the project. In that way, you can build habits of use early. If you wait for a crisis, then the lack of methods will probably aggravate the situation.

The following tools are appropriate for each area:

Category	Tool
Team communications	Electronic mail, groupware
Issue tracking	Network-based database management system, groupware
Lessons learned	Network-based database management system, groupware
Project plan information	Network-based project management software
Project documentation	Files on network server
Integrated project information	Groupware

The project management software allows members of the project team to update and add more detailed tasks.

Let's define more precisely the issues database approach. There should be three related databases. The first is the database for issues across all projects. This is employed as the master list of all issues. No issue can be identified unless it is a member of this database. The data elements are as follows:

Identifier of issue (index)
Issue title
Issue description
Type (category of issue)
General impact if not solved
Related issues
Related lessons learned
Date issue was first identified
The name of the person who created the issue
General actions for the issue
Comments

When you apply the issue to tasks in the project plan, you will identify specific elements and details of how that issue applies to the project. This is the second database, issues for the project. The data elements are as follows:

Identifier of issue (index)
Project identifier (index)
Status
Date issue was created for the project
Person who created issue
Status of issue (open, closed, replaced, and so forth)
Importance to the project
Specific impact of issue on the project, if not solved
Assigned to
Date resolved
Decisions made

How resolved
Related projects and tasks in schedule
Action items for resolution
Comments

To this you also have a third database to track the activity on the issue. These data elements are as follows:

Issue identifier (index)
Project identifier (index)
Date of entry (index)
Person making entry
Action taken
Result
Comment

The lessons learned database includes the following data elements:

Lessons learned identifier
Title of lessons learned
Content of lessons learned
Type of lesson learned
Situations to which the lesson learned is applicable
How to use the lesson learned
Anticipated results
Date created
Person creating lessons learned
Issues to which it applies
Projects and tasks to which it applies
Comments

As with issues, there is an additional file for comments on using the lessons learned. This file contains the following:

Lessons learned identifier
Date of comment
Person making comment
Comment
Suggested action

The lessons learned and issues databases should be applicable to all projects. In that way, you can assess and deal with issues and lessons learned that cross multiple projects.

Many project management software packages allow customizing of fields associated with each task. Here are some examples of the use of text fields:

- Person responsible for the task. This may be different than the resources used in the task. It supports accountability and aids in tracking (text field).
- Indicator of whether the task has high risk. The task is then management critical. This allows you to filter on only the high-risk tasks (flag—yes/np field).
- Issues related to the task (text field). This supports the linkage to the issues database.
- Lessons learned related to the task (text field). This is useful for linking with the lessons learned database.
- Related tasks in other projects (text field).
- Date task added. This is important for later analysis (date field).
- Reason task was added (for example, rework or revised estimate). This helps in your budget versus actual and planned versus actual analysis (text field).
- Organization unit responsible. (text field)
- Indicator of whether the task is in the template (flag field). This is useful for extracting a summary of the schedule for the template.
- Requirement that generated the task (text field). This field allows for tracking the additional work generated by the requirement—very useful in tracking the impact of changing or new requirements.

The project leader is responsible for establishing the initial data in each category on the network server for the project. This information can then be presented to the project team as it is employed during the project. Build a list of realistic situations or scenarios. For each compose a small plan and use playscript to clearly show each role in the method. Playscript is the technique whereby paper is divided by a vertical line into two columns. In the left column is the person or entity (computer) that is performing the step. The right column contains the detailed actions that are performed in that step. By doing this, team members will realize what specific tasks are expected of them. Using this organized approach, you seek as few areas of ambiguity and uncertainty as possible.

WHAT METHODS AND TOOLS WILL BE USED DIRECTLY IN THE PROJECT?

The methods and tools to be employed in the project depend, of course, on the nature of the project. Here is a potential list of areas in which methods should be identified. Note that these are identified for all projects to ensure consistency.

- Network
- Setup of software on a server
- Monitoring of software use

- Monitoring of network performance
- Troubleshooting
- Security management
- Antivirus controls
- Software development
- Languages and compilers
- Editors
- Debuggers
- Translators
- Object libraries
- Configuration management

Once you have identified the tools, you might think that you are finished. You are really just getting started. Here are some additional tasks for you:

- Determine the goals and expectations for each tool.
- Define methods for how each tool is to be employed.
- Identify contact points for methods and tools.
- Document methods and make them available on the network along with lessons learned and guidelines.
- Define how gaps between tools are to be filled.
- Establish rules for monitoring method and tool use.
- Define the parameters required to support the tools.

You will want to present your conclusions to the project team. Experience suggests that you give an overview for all areas. Then zoom in on those that will be employed in the next three months. Hold a similar meeting every few months.

How Do Templates, Issues, Lessons Learned, and the Plan Link?

Let's walk through the following diagram. The template is used to generate the plan. Estimating tasks and doing work in the project can use the lessons learned that are linked to the template. Individual issues are linked to the tasks in the plan. As you solve the issues, you may be able to extract more lessons learned. This, combined with experience, allows you to improve the template—resulting in cumulative improvement.

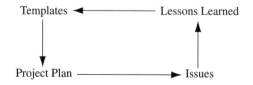

How Should You Develop Project Templates?

Recall that a project template is a high-level schedule. It includes summary-level tasks. Examples for specific projects will be given later. Figure 7.1 presents a simple template for upgrading a wide area network. A template is a more complete specification of a project than a work breakdown structure (WBS), which only includes tasks. The template includes resources, tasks, resources assigned to tasks, and dependencies between tasks. Templates allow you to analyze multiple projects and to prepare summary analysis reports for management. Without a template you will have to compose a summary schedule by hand from the disparate schedules. Templates offer the following additional benefits:

- Building, using, and maintaining templates give project leaders and others the chance to participate in the improvement of the project management process.
- Using a template to start with saves time.
- Junior project leaders can take advantage of experience through the lessons learned and the templates.
- Using a template can improve crebility of IT to the business unit or client.

To construct a template, follow these steps:

- *Step 1:* Create a general resource list. It will include general resources such as job titles (senior programmer or systems analyst, for example) as well as detailed resources (specific people and equipment). This resource list will be employed by both the templates and the detailed project plans and act as a resource pool.
- *Step 2:* Create a task list for the template in a top-down manner. Don't get down to individual tasks, as that is done in the creation of the specific plan.
- *Step 3:* Establish dependencies between the tasks in step 2. These are simple tail-to-head dependencies.
- *Step 4:* Associate general resources from step 1 to the tasks developed in step 2.
- *Step 5:* Link lessons learned to the tasks in the template.

Retrofitting an existing schedule to a template is covered later. To turn a template into a schedule for a specific project, you will add the detailed dates, tasks, and durations, and make the resources specific.

How Do You Establish an Effective and Flexible Task List?

Specific examples for major types of projects are included in Part III of the book. Follow the suggestions in constructing the template. You now have created the high-level summary tasks. To make this project specific, begin by adding more detail, but the added items still should be summary tasks.

ID	Name	Duration	Predecessors	Resource names
1	1000 Project Mgmt	2d		
2	1100 Setup Proj Plan	1d		Proj Ldr
3	1200 Establish Team	1d	2	Proj Ldr
4	2000 Reqmts Analysis	4d		
5	2100 Collect Data from Bus Units	1d		Sys Anal
6	2200 Collect Netwk Data	1d		Sys Anal
7	2300 Define New Rqmts	1d		Bus Unit,Proj Ldr,Netwk Anal
8	2400 Rev New Technol	1d		Netwk Anal,Netwk Mgr,Sys Anal,Proj Ldr
9	2500 Define Rqmts	1d	5,6,7,8	Bus Unit,Netwk Anal,Sys Anal
10	2600 Document Rqmts	1d	9	Sys Anal
11	2700 Rev Rqmts	1d	10	Bus Unit,Proj Ldr
12	3000 Sizing and Architecture	7d	4	
13	3100 LAN Sizing	1d		Netwk Anal,Netwk Mgr,Sys Anal
14	3200 WAN Sizing	1d	13	Netwk Anal,Netwk Mgr,Sys Anal
15	3300 Host Interfaces	1d		Netwk Anal,Sys Anal
16	3400 Software Spec	1d		Netwk Anal,Sys Anal,Vendor
17	3500 Prep Compon List	1d	13,14,15,16	Netwk Anal,Sys Anal,Vendor
18	3600 Determine Costs/Benefits	1d	17	Bus Unit,Sys Anal,Vendor
19	3700 Cost/Benefit Analysis	1d	18	Sys Anal,Bus Unit,Netwk Mgr,Proj Ldr
20	3800 Document Analysis	1d	19	Sys Anal,Proj Ldr
21	3900 Review with Mgmt	1d	20	Proj Ldr,Bus Unit
22	4000 Procurement	4d	12	
23	4100 Update Plan	1d		Proj Ldr
24	4200 Prepare Spec	1d	23	Netwk Anal,Sys Anal
25	4300 Setup Procur w/ Purchas	1d	24	Proj Ldr,Sys Anal,Purchasing
26	4400 Procure Hardware	1d	25	Purchasing,Netwk Anal,Proj Ldr,Sys Anal,Vendor
27	4500 Procure Netwk Comp	1d	25	Purchasing,Netwk Anal,Proj Ldr,Sys Anal,Vendor
28	4600 Procure Software	1d	25	Purchasing,Netwk Anal,Proj Ldr,Sys Anal,Vendor
29	5000 Installation	3d	22	
30	5100 Cabling	1d		Installers,Netwk Anal
31	5200 Netwk Hardware	1d		Vendor,Netwk Anal
32	5300 Hardware	1d		Vendor,Netwk Anal
33	5400 Netwk Oper Sys	1d	32	Vendor,Netwk Anal,Netwk Mgr
34	5500 Sys Softwre	1d	30,31,33	Vendor,Netwk Anal,Netwk Mgr
35	5600 Netwk Mgmt	1d		Netwk Anal,Netwk Mgr
36	5700 Est Netwk Segmts	1d	35	Netwk Anal
37	6000 Testing/Production	5d	29	
38	6100 LAN Testing	1d		Netwk Anal
39	6200 WAN Testing	1d	38	Netwk Anal
40	6300 Initial Product	1d	39	Netwk Anal,Netwk Mgr
41	6400 Post Imple Measuremt	1d	40	Proj Ldr,Sys Anal
42	6500 Post Implement Rpt	1d	41	Proj Ldr,Sys Anal

Figure 7.1 Template for Expansion of Wide Area Network

It is assumed that the team has been identified. If not, then you will have to complete these steps yourself:

- *Step 1:* Identify task areas for each team member. In your plan, you can use the text field for the person responsible to indicate this.
- *Step 2:* With a summary schedule, sit down with each person and review the plan. Have each participant attempt to create detailed tasks and dependencies.
- *Step 3:* Review what each person does and refine it.
- *Step 4:* Collect what each person does and standardize the wording and abbreviations of the tasks.
- *Step 5:* Assign other resources to the tasks.
- *Step 6:* Establish linkages with other projects.
- *Step 7:* Fill in the durations and dates.

How Should You Assign Resources?

With the task list and general resources, you can replace general resources by specific resources. For example, the resource "systems analyst" could be replaced by the person's name. You can also assign specific resources to each detailed task. Systems projects usually have only a few key resources for each task. Don't include resources that are commonly available as this will just create more work for you.

How Do You Relate Areas of Risk to the Plan?

Recall that risk management relates to the management of issues. Each team member identifies tasks he or she feels have substantial risk. The risk can relate to the technical nature of the work, the team member's level of knowledge and experience, uncertainty about what is involved in the task, and relationship to other work. Indicate risk in the appropriate text field of the plan. You are now in a position to assess risk by relating known issues to tasks with high risk. You may encounter three potential situations here: (1) If you have an issue with no related tasks, then the task list is not complete. (2) If the issue relates to tasks that are not yet labeled as risky, then consider relabeling. (3) If you have high risk tasks with no issue, then you are missing issues. This analysis helps make the plan more complete and integrated.

This method is different than putting in contingencies. Some people pad their task durations to be safe. If you do that, you hide the risk and issues. The schedule will also be too long and unacceptable. It is better to identify issues associated with the tasks that you cannot estimate and, if necessary, add additional tasks that relate to the contingency. Hopefully, these tasks will never have to be performed.

How Should You Establish Dates and Durations?

Have team members estimate durations of all detailed tasks. If the estimate exceeds 10 business days, then define even more detailed tasks. If the tasks are all constrained to begin as soon as possible, then there will be few dates to enter. Dates will be set by dependencies, the start date of the project, and durations. For flexibility in planning, maintain the scheduling constraint "as soon as possible." If you fix the dates for a task or change the constraint to "as late as possible," "must start by," "must end by," and so on, then you lose flexibility. You will often get error messages because some tasks will have negative slack.

What happens if you have trouble estimating durations? Chop up the task into smaller parts. Then if you still cannot do the estimation, identify the subtask that you cannot estimate. Why can't you estimate it? The reason is an issue that you can add to the issues database. Work on the issue with others to define an estimate. It is positive that you define these issues early. Don't succumb to the temptation to just put down some estimate!

It is possible that the schedule produced will not be acceptable. The end date will be beyond that defined in the project concept. If this happens, go back to team members and first review the detailed tasks for the next three to four months and make adjustments here. Next, move to future tasks. Test the durations and dependencies to try and move the dates in. Remember that you, as the project leader, are being tested by the team members. If you give in to their estimates of durations too easily, then they may assume that you can be pushed around. You will have more problems later. Here are some additional tips on compressing the schedule. See if you can do more work in parallel. This may require you to divide tasks into subtasks. Another suggestion is to work on the issues behind the tasks that have risk. This is better than trying to compress tasks that have no risk.

Here are some tips for dealing with team members:

- Probe the reasons behind their estimates. You want to get at their assumptions.
- Try to break up the task into subtasks. See if these can be performed in parallel.
- Do what-if analysis in their presence to determine the impact of changes.

How Should You Link Projects?

To just say that two projects are linked is almost useless. You cannot perform any analysis. Carry out the following activities to establish linkage:

- Link projects at the detailed task level as well as at the summary level using the project management software. This is possible because all of the plans are on the network server.

- Identify the individuals on each project team that will be responsible for interfaces between the two projects.
- Identify a list of potential interface issues.
- Establish an interface method for addressing issues with the managers of the related projects.

How Do You Create the Project Budget?

With a schedule defined, you can work on estimating the budget. Experience for systems and technology projects tends to favor an approach where you start from the areas that are the most defined and have the least risk and then proceed to the areas where there is greater uncertainty. A typical project is composed of hardware, network, software, and personnel costs related to development and implementation. Some people make a big deal out of hardware cost estimation. This is typically based on the sizing of the hardware required. However, hardware takes up a smaller percentage of a project budget each year and is one of the most certain costs. The network has greater uncertainty because there may be problems in implementation and additional network software that was not known at the start of the project. Software packages can be estimated more easily in terms of cost. The problem is that you might miss some elements of software. The greatest uncertainty and risk lie in personnel time and effort. Consider the following guidelines for estimation.

Hardware

Estimate hardware first. In terms of end-user hardware, allow for a fully configured system and more workstations or PCs than you require to allow for change. Lessons learned indicate that people from the business unit will underestimate their hardware needs. Then when they see the system, they will change or increase their requirements. Perhaps, make the end-user hardware an item separate from the project budget.

With respect to the server hardware, this depends on the specific company and situation. In the past, underestimation has been due to not including history data for on-line access, databases for ad hoc inquiry and reporting, and data warehousing type applications. In terms of labor associated with hardware, ensure that you have covered testing and installation.

Network

For cabling connect up the reasonable maximum number of locations. This is because the greatest single cost factor is the labor involved in installation rather than the cable itself. To have the installers return and cable additional offices can be very expensive. Network hardware and embedded network software are straightforward if you have one or only a few vendors. In your budgeting, allow for the hardware to be fully featured, even if you actually order less. If you are adding a new application on top of an existing network, then consider adding money for hardware upgrades due to the increased workload.

The network management, diagnostic, and control software is more complex because it may operate on host hardware and may involve additional staff. Try to place a limit on the budget amount by setting criteria for turnover to production and operations as well as on the labor needed for installation.

Software

The system software is easiest to estimate and include. Utilities are often underestimated in that people miss some of the software required. Talk to the vendors as well as other software users to determine what software they have acquired. Even if you do not elect to purchase it now, put it in the budget. Next, try to apportion the software cost between ongoing operations and the project. It is not really fair for the project to pick up operational and maintenance costs. For software tools such as development environments, database management, fourth-generation languages, and so on, some of the leading factors in underestimation are (1) not including all of the necessary modules, (2) not allowing for sufficient training in the software, and (3) failure to include vendor consultations to get you started.

The preceding remarks apply to software packages. Additional factors here are the customization required of the application software to fit specific business requirements and the design, programming, and the testing of interfaces with other software. Solicit vendor estimates as well as experience from existing users of the software package. With major integrated software packages, you can seriously underestimate the costs.

Staffing

With the schedule and resources identified, you can determine the people and their costs. Then you can add overhead. For contractors you can solicit specific bids for task areas. If this sounds so straightforward, then why does it go wrong? What happens is that requirements or technology changes. The schedule then slips. Once the schedule slips, the labor costs go up. The costs are a trailing indicator of the problem. The core of the issue is how to protect the project from change. If there are changes, then an allowance has been made for changing both the schedule and the budget. Estimation gets a bum rap here because somehow people are expected to see into the future. However, there is no extrasensory perception for software development yet.

What should you do if you do a good job estimating and end up with a number that is too large to be acceptable? Do you arbitrarily cut the number to fit someone's plan? The answer is no. Try to move some of the costs out of the project and into the business unit or other areas of systems. The project budget may look big to information systems, but it typically is much less significant to the business units—and, after all, they are ones who are going to benefit from the new system and technology. For PCs and the network, you can argue that the business unit will eventually require these items to complete its standard tasks, not including the new system. The new system represents an incremental application on the network and hardware.

Some of the common mistakes people make are the following:

- They fail to include sufficient hardware for development, test, and production environments. These should be three separate environments. Otherwise, you are likely to have incomplete testing and quality assurance.
- There is software missing. When you acquire a substantial fourth-generation language, for example, there are many additional components and utilities that you should get. Validate the software components.
- If the project is going to go for a year, then you should allow for upgrading to new versions. Many people only put in the initial purchase or acquisition cost.

Moving on to staffing, go to the project template and the estimate of the plan. In Microsoft Project and other software project management tools, you can obtain a "Resource Usage" view of the data. Here the rows are the resources and tasks and the columns are time periods. The entries in the table constitute the number of hours that the resources work during that period on the tasks. Export this table to a spreadsheet. There you can apply money to the resources. Don't use this as is. Follow these guidelines in doing the analysis.

- Allow for additional resources that are not included in the plan. After all, the plan is not a detailed budget entity. Rather, it is a management one. If you attempt to make any project plan totally accurate in resource allocation for each and every small resource, you will not have enough time to update the schedule. It will be too unwieldy to use.
- For consultants and contractors you should allow for startup and termination.
- Remember that the project plan is for work in the project. There are also support tasks that have to be included in the resource estimates.
- A rule of thumb is to take the total number of hours in a given period such as a month and then apply an average cost. This will give a ballpark starting number.

How Do You Determine Benefits?

Benefits were discussed earlier in general. Now you are going to define these benefits in more concrete terms. Here are some guidelines on benefits:

- Insist that all benefits be tangible. Management has heard enough fuzzy benefits.
- Insist that the business units develop the benefits with your help. They should stand up before management and present these. If you are a consultant or in IT, you are just not credible because you do not know the business and, more important, you cannot make these benefits materialize.
- Determine how the benefits will be achieved.
- Pin down how the benefits will be measured. This is best done by measuring the business processes that are impacted before and after implementation.
- Benefits should be revenue or cost oriented. Any cost reductions should be supported by how labor will be cut or reallocated.

These are tough guidelines, but they are essential for credibility and support.

How Should You Establish Multiple Projects?

Most of the techniques presented for single projects apply to multiple projects as well. Here are some additional guidelines:

- Build high-level tasks for all projects first.
- Establish dependencies between projects at this high level.
- Identify common issues and associate them with appropriate tasks.

It is important for the line and project managers involved to build their plans in a collaborative, shared information environment. Otherwise, it will be more difficult to resolve issues. Information hiding at the start of the project can really hurt later and result in more political games and maneuvering.

How Do You Establish a Baseline Plan?

Most project management software allows for multiple sets of start and finish dates for each task. When you are setting up the plan, you are typically working with one set of dates—scheduled dates. You can save these into another set of dates and then make changes to see what happens to the schedule for what-if analysis. The software allows you to set the baseline plan dates. This typically means that the scheduled dates are saved as planned or baseline dates. This is referred to as setting the plan. Later during the project you will enter tasks as complete and make changes. These will be saved from the scheduled dates into the actual dates.

How can you take advantage of the other dates in systems projects? Some examples from lessons learned are the following:

- *Customer or business unit dates.* These may be separate or in addition to the project manager dates.
- *Modifications to baseline.* Due to issues, resource reallocation, and so on, a major schedule change may be necessary. You can still keep the baseline dates by saving them into another set of dates. Then you can establish the new baseline plan. The original baseline plan is then not lost and can be employed for later analysis.

You must retain any significant change to the agreed upon schedule due to requirements changes, external factors, and so on. You can then track and present the history of the project. Most software allows you to show three sets of dates in the bars of the GANTT chart. For example, in Microsoft Project, you can store many different start and finish dates. If you run out of these, then you can save the schedule to retain the older dates.

HOW DO YOU SELL THE PLAN?

Let's consider each audience you are attempting to reach. Your objective and approach will be different for each.

- **Systems and upper management.**
  The plan must mesh with the project concept. If there are major differences in dates and budgets, then you will have to redo the concept to fit the plan and market it prior to presenting the plan. Now put yourself in management's shoes. What do you ask about a plan? Here are some examples:
  —The project concept identified initial ideas about risks. What has the project plan done to minimize the risk?
  —What resources does the plan require? Are these available? Will these demands affect other current work?
  —Are interdependencies among projects clearly identified? How will they be managed?
  How do you sell the plan to management? Review the project concept and highlight how you have addressed risk. Next, discuss how the plan and project will be managed in terms of the business unit role and interfaces with other projects. Another area to discuss is resource management. How will the team be coordinated and managed?

- **Project team.**
You will succeed in marketing to the project team if team members participate in the updating of the plan and in project management as well as in
giving support for the schedule dates. You fail if all that you achieve is
understanding.

    Go over the plan with all members of the core team and major part-time
team members individually. Review their tasks and dates. Have them identify risks and more detailed tasks for their work. After doing this, gather the
team together and present the plan along the lines of management presentation discussed earlier.

- **Business units.**
There are specific marketing objectives here. The plan reinforces and defines
the role of the business unit. You want the unit to acknowledge and accept
this role. The business unit must see what responsibilities its employees will
have and how important they are to the project. Also, you want people in the
business unit to understand that the plan can help in communications and in
resolving issues.

    Visit the business unit managers and review the project concept. Present
the plan at a summary level. Don't bury them in detail. With this overview
show them what tasks their staff will be performing, when this will occur,
and the importance of the work. How do you indicate importance? One way
is to show how slippage will occur if the tasks are not performed on time.
Another is to explain the impact of not completing a task.

    Many project management books end here. However, you must set the
stage for addressing questions and issues with the team later on. From
lessons learned and the initial issues identified in the project concept, discuss
some of the likely issues that will be faced. This can naturally lead to a discussion of the issue resolution process and how the plan and issues will be
updated. Return to the issues and select one. Go through the process of issue
analysis and resolution. An example of an issue is what will occur if the person assigned to the project is no longer available.

How Do You Fix an Existing Schedule?

Here is a step-by-step repair process:

- *Step 1:* Review the resources assigned to tasks and establish a standard list.
This means that you will gather all resources across schedules and compose
a standard list. Then you will edit the list of resources in each schedule to
conform—thereby ensuring compatibility.
- *Step 2:* Establish the highest level summary tasks in the schedule. Create a
top-level summary schedule of fewer than 20 summary tasks.

- *Step 3:* At each level introduce dependencies between tasks and insert dependencies that are tail to head.
- *Step 4:* Move down in the level of detail to lower summary tasks. Keep adding more detail, and then add dependencies at that level.
- *Step 5:* When you have fixed the detailed tasks, add the resources from the resource list to the detailed tasks. You should assign no more than three to four resources per task at the detailed level.
- *Step 6:* Enter dates and durations of the detailed tasks. These will normally be the scheduled start dates and durations for the detailed tasks.

EXAMPLE: ELECTRONIC COMMERCE

Let's give an example of a template for electronic commerce. Once a company embarks on electronic commerce, it is likely to have several electronic commerce projects for individual transactions, justifying the creation of a template. Key areas of the project relate to business partners, interfaces to legacy and other host systems, and controls.

1000	Project management
2000	Business partner relationships
3000	Business and process analysis
4000	Technology support
5000	Host/legacy system interfaces
6000	Security, firewall, and network access
7000	Software development and implementation
8000	Electronic commerce controls and security
9000	Technical integration and testing
10000	Process changes and procedures
11000	Marketing changes and support

Note that the areas have been employed to define the project structure by highlighting integration and marketing. Who might be involved in the project? Beyond IT you have the business units including marketing.

If you were already doing electronic commerce, you would probably have already completed a small project to define strategy, costs, benefits, assessment of competition and Web sites, and overall schedule as well as technical direction. This standardization can then be employed across other electronic commerce projects.

E-BUSINESS LESSONS LEARNED

Building the plan for e-business requires a great deal of collaborative effort. Estimating the hardware, network, and software is easier than the business side. Here are some guidelines.

- Make sure that you include a test environment for the software. E-business requires you to be thorough in quality assurance. In one company there were several errors that were undetected. The software was put into production. Someone found out that they could order and receive merchandise free by combining promotions and discounts, and this information was posted in a chat room. Within a week, the firm had unknowingly given away over two million dollars in merchandise!
- Use external information as a start in estimating benefits. This will identify the areas of benefits for implementation.
- Be sure to include the benefits of changed internal business processes for non e-business work. In many cases, this is where the tangible benefits will be initially received.

GUIDELINES

- Keep task descriptions simple and use a standard set of abbreviations (for example, Rev for review). Number all tasks for easier reference later. Keep task descriptions short (25–35 characters) so that the GANTT chart will still fit on one piece of paper.
- Set the font to be Times New Roman 12 point. This is standard and easier to read than smaller fonts.
- Review all task descriptions. Split tasks that are complex sentences.
- Each task should begin with an action verb (define, develop, review, collect, design, and so forth). Milestones indicate what is to be completed (such as prototype completed or tested). This ensures consistency and differentiates between tasks and milestones.
- If a team member has never done task planning before, work with him or her to define tasks and get started. Your goal is to have people be able to work alone, but also work together. If you don't accomplish this goal, you could be locked into the old mold where the project leader gathers the status information and becomes a clerk.
- To market the idea of a template, develop a strawman template and have people review and comment on it.
- Make sure that you develop the plan areas with the business unit staff as well as internal systems team members. Have them begin to enter tasks into the project management software.

WHAT TO DO NEXT

1. Collect several project plans and review the task structure for the plans. Answer the following questions for each plan:

- Are the high level, summary tasks consistent in the schedule?
- Does the project cover all of the areas of risk?
- Is the level of detail consistent across the schedule?
- How have updates and changes been made to the schedule?
- Is it easy to find specific tasks or do many have the same names?
- If you had an issue in the project, then could you find the associated and affected tasks?
- Has a baseline schedule be set?
- Is the wording of the tasks consistent?

2. Move up from single to multiple projects. Answer the following questions.
   - How easy or difficult is it to create an overall project plan that combines several project plans?
   - What inconsistencies and problems have to be fixed first before you can combine the projects?
   - Is it easy to find issues that cross projects?

3. Assuming that you were going to build project templates, how many and what different templates would have to be constructed? What types of resistance would you encounter?

SUMMARY

Developing the project plan is a political process in which you reinforce how people work together, what they can expect, what is expected of them, and their commitment to the project. If you miss this opportunity, then you will face an uphill battle in establishing methods later.

By empowering the team, the project leader can spend more time on issues and on doing project work. The team members can think on a larger scale about how to do their work and the impact their work will have on the overall schedule as opposed to just focusing on immediate tasks. This collective approach also fosters more joint work among team members.

Part III

Managing Projects

Chapter 8

Effective Project Tracking and Coordination

INTRODUCTION

In the past, much of the emphasis in managing a project was directed toward updating and analyzing the schedule. Today, in systems projects it is a different situation. Updating and analyzing are still important, but now they are trailing activities. Problems and issues have already surfaced by this time. It may be too late to prevent problems if you wait until slippage occurs.

Adopt a more proactive approach in managing projects. A primary duty of the project manager is to identify, analyze, and resolve project issues. An issue can be a problem or an opportunity. If you address issues, then you can reduce risk and increase the chance of project success. This is such an important activity that the chapters in Part IV address more than 60 specific issues in different areas. Alternatively, if you let issues linger and concentrate on project accounting, you jeopardize the project.

A second part of being proactive is to participate in some of the project work and to review work quality. This can extend across several projects. If multiple, related projects are not managed together, then each project can be successful individually, but together there is failure caused by a lack of coordination.

It is necessary to define your goal in managing and completing projects. Traditionally, systems projects were successful if they were completed on time and within budget. Today, this is unacceptable. You must achieve the benefits from implementation and completion of the project. This is asking for more, but it is consistent with the earlier discussion of the project concept. Failure then lies in not only not completing the project, but also in not achieving the benefits. In response, you can say that getting the benefits is not the responsibility of the systems group. This is true and points to the need for a sufficiently wide scope to accommodate process change and improvement. You want the tasks of measurement and implementation of process change to be included in the project plan and performed by the business unit. This will result in a longer project time line. Another impact of this definition is that business unit management and staff will be more heavily involved in the project.

APPROACH

HOW DO YOU IDENTIFY AND MANAGE ISSUES?

A project leader must constantly be looking for potential issues—both problems and opportunities that can impact a project. Issues can surface internally from team members or externally from other projects or groups, business units, vendors, and management. That is one reason why a good project manager spends much of his or her time communicating with these entities. The last thing you want is to be caught off guard by a nasty telephone call, memo, or e-mail message.

Once you have detected something, you have to understand it. Is it a symptom of an existing, known issue? Is it new? Here are some questions to ask. A symptom is an effect of an issue.

- Is it a symptom or an issue? If you fix it and a problem remains, then it is an issue. If it is a symptom, create a statement of the underlying issue.
- Is it related to an existing issue? If so, is it another symptom? Does it expand the issue? Can it be treated separately?
- What are other symptoms and potential causes of the issue?
- What are the impacts of the issue on the project? On other projects? On the organization?
- What is the priority of the issue?
- What happens if the issue is not addressed?
- What are related issues?
- Who will analyze the issue?
- What will be the analysis approach?

Let's give an example. Suppose that a team member tells you that the business unit wants to change requirements. There is a tendency to run over to the business unit and determine if it is true and what the new requirement is. You can see how the requirement can be addressed. This knee-jerk reaction is often counterproductive. You risk being "nickel and dimed" in changes. Instead, treat the issue through analysis.

Turning to analysis, first, determine the status of work in the project. This will provide you with perspective later when you meet with the business unit. Second, assess relations and communications between the team and the business unit. Why didn't this change surface before? Why does it appear now? Next, plan for contact with the business unit. Push for a review of all requirements as well as project status. This will put you in a proactive position as opposed to being on the defensive.

In the meeting with the business unit, relate each and every change to the business process. Get the tangible benefits out on the table. Test what happens if the change is either not made or made through manual procedures. If the change remains real, then you have a bonafide issue on your hands that you can address.

Establish an issues database as was discussed earlier. Update the database with new information on a regular basis (at least once or twice per week). Using this database you can now move to group related issues. Seek to solve multiple issues with one set of decisions and actions. Return to the example. Let's assume that there are additional issues related to the business issue and commitment. For example, suppose that the business unit is not actively participating in the project to the extent originally agreed. Because you are dealing with requirements already, it is natural to include these as well. There is the potential for a trade-off here— you support the new requirement; in return, the business unit provides more support to the project.

How Do You Measure Open Issues?

Whenever you have a number of open issues that vary in priority, impact, and age, you can create a graph such as that in Figure 8.1. The chart tracks the number of open issues by age for different priorities and in total. The curves have been smoothed. This chart is normal and desirable because it is logical that most of the open issues have been discovered recently.

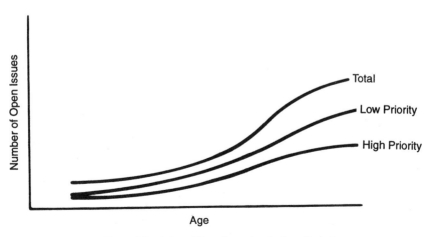

Figure 8.1 Aging of Open Issues in a Systems Project

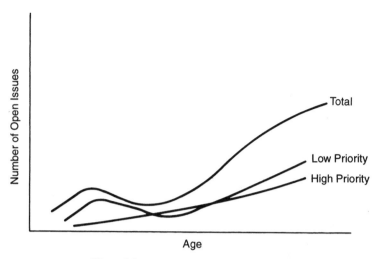

Figure 8.2   Problem Situation of Open Issues

The graph in Figure 8.2 is more serious. Here substantial high-priority issues have remained unresolved for some time. Yet the total number of issues remains similar, as does the lower-priority issues. How does this happen? Political or financial problems may be associated with these older issues, which management chose not to address. It can also be the fault of the project leader due to inexperience or the lack of will to address the issue. Yet these issues fester and can seriously impact the project depending on the decision. Look at the changing requirements example. Suppose the decision is deferred. Then if the final decision is made to implement the requirement, much of the project work may have to be redone.

WHAT GENERAL ISSUES ANALYSIS DO YOU PERFORM?

Let's consider how you should analyze issues overall. Start with these basic data elements.

- Issue number
- Status
- Type
- Date discovered
- Date resolved

These simple data elements can give you the basis for many comparative graphs. Here are some examples:

- Total number of issues at point in time. What is the trend in the total?
- Total number of issues by type. What is the trend in business, technology, external, and process issues?
- All issues by date discovered. Number of issues by discovery date. Are more issues being uncovered later in the project?
- All issues by date discovered and type. Number of issue by discovery date for a specific type. Are more issues of the specific type being uncovered?
- Average time to close issues by date of discovery. Is there a trend as to how long it is taking to resolve issues?
- Average time to close issues by type and by date of discovery. Is there a trend as to how long it takes to close issues of a certain type?

How Do You Measure the Work in a Systems and Technology Project?

Measuring the work starts with collecting information on status. If the project team has been updating its part of the schedule and issues, then you can start with a review of this information. If collaborative scheduling is not in place, you have to determine status manually. Don't fall into the trap of holding a project meeting to determine status. Instead, meet with each person in his or her office if possible. Plan ahead for the sequencing of the meetings with team members. Start with the team members who are working on tasks with the greatest risk. This is probably where you can do the most good.

Here are some guidelines for individual one-on-one meetings:

- Ask how the work is going and what issues or problems the team members have run into. This will get team members to open up to you in their own words.
- Translate what they say into the impact on the tasks they are working on. Have them review what you did. This is more effective than just handing them a list of tasks and asking them to check off what is complete.
- For each task that is completed, plan on how you will review it for quality.
- Now go down the list of open issues and determine which are impacting their work. Examples might be availability of resources or waiting for other team members to complete their work.
- During this meeting, ascertain whether they are communicating with other team members on their own. Or are they working alone? If they are working alone, plan ahead for a small meeting to get them together with other team members.
- Ask what resources or help they need for current and near-term future tasks. Volunteer to get them help.

- Don't ask what percent complete their current task is. This is a very ineffective question in software and technology projects. Instead, ask them what is left to do in the task. This focuses on getting it done as opposed to how much they have done.
- When addressing future tasks, ask them to define the work in detail. This will help flesh out the detailed schedule. Of course, it is better if they maintain their own tasks because they can be self-sufficient and more involved in project management.

You can review work at several levels—just like child's homework. You can take the team member's word for it. You can ask to see it, but you don't review it. Or you can have the work reviewed. Time doesn't permit a full review of every end product. Moreover, in systems projects a number of milestones are almost self-evaluating—that is, the work is immediately put to use and tested. Other milestones might not be critical at the time. Several criteria are applied for picking which milestones or end products to evaluate:

- *The most important milestones.* This is traditional project management. It makes sense if you can know which are the most important. However, if people know which milestones are important, they will consciously do well on these.
- *The milestones that have the greatest risk.* These are the ones on which much later work depends. It is recommended that you use risk as the main criterion for selecting milestones for review. In some cases, risk and importance lead to the same milestones.

How will you evaluate a significant milestone? Employ an approach that uses several people from the team. You want to use the review to build the team dynamics. Have each person involved in the review develop a list of questions ahead of time, then ask how he or she want to do the review. The purpose is to encourage a proactive, positive review process.

What do you do after a meeting? Don't wait until you talk to the entire team. If specific actions are necessary, such as follow-up on procurement or installation, get on this right away. Report back to the team member. Part of your job as project leader is to identify and address issues.

After you have determined status, update the issues database first. Then go to the project plan and schedule. Here are some guidelines for updating the schedule:

- Mark all of the tasks that are complete first. Don't use percent complete for detailed tasks.
- Insert more detailed tasks for current and future work based on the information that the team members provided. Insert the date that the task was created in a text field within the project management software. Use some symbol to indicate that it is a new task.

- Mark the tasks that are now started.
- Associate the issues with the tasks for traceability.

Do all of this with a copy of the schedule. Now filter or extract the tasks for each person for the last month and a future month. Have each person on the team review what you have done. Make any corrections and changes to the schedule. Update the schedule. This will become the new actual schedule. This is the manual version of the process. In collaborative scheduling, each person performs the update steps listed. What are some of the problems that you can face when reviewing work?

- *The work is behind schedule.* Reasons for this are (1) the original estimate was too optimistic, (2) the resources were not available to do the work, and (3) the scope of the task was not understood properly.
- *The quality of the work is not acceptable or complete.* This can be due to a misunderstanding of the task. Another reason is that the person doing the work lacks specific training. That is why it is urged that each person define his or her own tasks.

What do you do? Some people might just extend the schedule. Don't do this. Create an issue and follow up. If the person doesn't appear to be making progress, volunteer to get help. Many people try hard and because of ego or some other reason don't want to admit problems. Don't blame them. Get them help. This is especially true when you have a mix of junior and senior programmers and analysts. Encourage the senior team members to help the junior team members. Emphasize to the senior staff that part of their job on the project is to mentor and help the junior members.

How do you handle these situations in the schedule? Instead of extending a task, create a new task next to it and indicate the reason for the change (such as rework or scope). Leave the old and new tasks open and still being worked on. If you have task 8110 Design update program, then you would add task 8111 Complete design of update program.

How Do You Analyze a Project?

You now have updated the schedule. You can produce several of the standard graphs and reports, or you can produce an updated GANTT chart that shows actual versus planned results. However, before you do this, perform an analysis of the schedule. Here are some specific steps for analyzing the schedule:

- Look at the number of outstanding issues and plot a graph such as that presented earlier.
- Examine the oldest outstanding major open issues. Why are these still open? Start doing research to determine what is going on. What can you do to resolve these issues?

- With the update you probably have added more detailed tasks. The schedule probably has now been extended. Don't leave this as is. Look at the future tasks and find those that are likely to require more detail. This will help you see how bad the situation might become.
- Look at the future tasks by person and determine which have high risk. Label these in a field in the project management software.
- Filter the schedule by extracting those future tasks with high risk. Has the number and extent of risk grown? Are the tasks with risk on or near to impacting the mathematical critical path? Here you are working with the management critical path.
- Now you can copy the current schedule and create a model of what the future might hold. Adjust the task durations based on the previous analysis steps. This will probably be scary because it will extend the schedule even further in the future.
- Review the list of issues and compare it to the future tasks that have high risk. There are several possibilities here. For tasks that are risky with identified issues, assign these issues high priority. For tasks that have risk but no issues, you must attempt to identify the issues that lead to the risk. You can consult the team members who will be working on these tasks in the future.

Do you realize what you are doing in this analysis? First, you are being proactive in addressing issues with current tasks. Second, just like team members who prepare for the future tasks, you are preparing for the tasks that have high risk. You are attempting to identify and address issues in these tasks so that the risk can be reduced. It is a myth that you have to wait until the task starts before you define the issue. Don't wait that long.

At this point in your analysis you have several schedules. You have the updated current schedule with the additional tasks and revised estimates from the project team. In addition, you have the baseline or plan schedule. Remember, this is the one that management sees as official. You have also created a third schedule based on future estimates. It makes sense to put these together in a GANTT chart. However, the baseline has less detail than the current schedule which, in turn, has less detail than the "future" schedule. How do you put these together? Summarize each schedule to the same tasks. That is, roll up the tasks to the summary level. The summary level is the same for all three schedules. In some project management software packages, you can now combine these schedules so that each summary task has three sets of start and finish dates in the GANTT chart. An alternative approach is to create a new summary schedule and copy and paste the actual, planned, and future dates into the new schedule. Three sets of dates is normally the maximum number that can be shown on a GANTT chart clearly.

The GANTT chart can now be employed to do further analysis. You can answer the following questions:

- What is the impact of the current information on the current status of the project? If there is little or no change, then no action is required. If there has been a major change in the schedule, then proceed to the next question.
- What is the estimated effect of current knowledge on the future schedule? This will indicate whether there is major slippage in the future.

There are now several cases to consider. If both the current and future do not change the scheduled completion, then no immediate action is needed except for the actions identified earlier. If the current schedule is okay and the future schedule shows slippage, then you must take several actions. First, you must follow up on the issues related to future tasks to reduce risk and possibly improve dates. Second, alert management informally of the potential for slippage and explain what you are going to do.

In the third case, the current schedule slows slippage but the future schedule shows either improvement or the same slippage. What you do here depends on the degree of slippage. If it is severe, then you may want to follow the steps in the last case that follows. If it is not severe, then you can report that to management along with actions you are taking.

In the fourth case, there is slippage in both the current and future schedules. Now you must do more analysis of the issues to verify that your estimates of the future are correct based on your current information. If they are or when they are, you can proceed with considering alternatives. Do this analysis prior to any management presentation. Here are some analysis steps to consider:

- Can the structure of the project be changed to have more tasks performed in parallel? This what-if analysis can be performed with your future schedule. Save it before you start. Save the revised schedule with the structure change.
- Go to the list of open issues. See which ones impact the schedule. Revise the future schedule based on the structure change in the preceding step with changes to show the effects of early and favorable outcomes of issues. Save this schedule separately as "Issues."
- What happens if more nonpersonnel resources were added? This might include additional hardware or software. In other words, what happens if you throw money at the project? Do this analysis with a copy of the schedule with the structure changes. Save this schedule as "Resources."
- What happens if you add more people or add more time from current staff on the project? The personnel can be either employees or contract people such as contract programmers or consultants. Save this schedule as "Personnel."

You now have four different models of the future. For each of these schedule versions, prepare an estimate of the budget impact so that you have a budget-versus-actual analysis. These models can now be discussed with management.

This analysis takes considerable time. It is recommended that at most you do this once every few weeks or month. Otherwise, you will spend all of your time doing this analysis. You can reduce the time needed for this analysis by being very organized in your files and work habits. Here are some guidelines and lessons learned:

- Do the schedule update at a different time from the analysis. Updating is mechanical. Analysis is not. Doing them together will also take too much dedicated time and requires different attitudes.
- Organize your computer files so that the updated and baseline schedule are on the network. The additional files you have created are on your hard disk. For each schedule you create, document the assumptions and changes that have been made. Otherwise, you will not be able to remember all of them and will have to reconstruct your work, which can be very time consuming.
- Print out each GANTT chart and label it. Put these charts in paper files along with labels as to assumptions and dates.
- Summarize your analysis results in bullets organized as follows:
  —Assumptions
  —How the schedule changed
  —Impact of schedule change
  —Feasibility of making the changes

If you are considering changing the schedule after meeting with management, make sure that you include future changes as well as changes based on current information. The goal in systems projects is to have as *few major changes to a schedule as possible.* Managers will question what you are doing if you keep changing the schedule and running to them with problems. Having constantly changing schedules will likely disturb the project team. Team members will question the direction of the project.

How Do You Perform Budget versus Actual Analysis?

You can track the expenditures for equipment, the network, and software in standard ways. For labor analysis, you would use the same approach as during planning. You would generate the "Resource Usage" view of the data of the project. This would be then be exported to a spreadsheet. There you can compare it to the same information for the baseline plan. Once you find discrepancies, then you can zero in on the tasks that created the differences.

How Do You Track Multiple Projects?

What do you do differently with multiple projects? Begin with analysis of the schedule for each project. If you have project leaders for these projects, have them perform the preceding analysis. Work with them individually to review their work. Identify problem areas and issues for each project. Work on the issues that cross multiple projects and prioritize the issues across projects. An example of an issue might be the need for specific programming software tools. Another example is the need for additional design or programming resources, which have the same area of expertise. This will move you up to the level of perspective across multiple projects. For each common issue do the following:

- Formulate a general version of the issue that will apply to each project.
- Define what will happen to all projects if the issue is not resolved, and do this by showing the impact of slippage on the schedules.
- Define potential solutions for the issue and show how these impact the schedule.
- Estimate the cost and benefit of the potential solutions.
- Determine how you and the project leaders will control and manage the solution to ensure a result.

Now download the project schedules that the project leaders have developed from the network onto your PC. Filter or extract the tasks from each schedule that are currently active or that occur in the next month or two. Combine these extracted schedules into a single schedule. You can now see the resource requirements across each schedule. This only works if the project leaders use the same set of names for resources (for example, they use the same resource pool), they follow the same general template forms, and they have assigned resources to tasks. You can now perform your own what-if analysis to see what happens if you reallocate resources. As you make changes that appear to improve the schedule, make notes as to the changes that were required.

With this analysis in hand, the next action is to gather the project leaders in a meeting and see how resources can be shared across the projects more effectively. Use a PC attached to a video projector so that you can do the same what-if analysis as a group. Here you can show them the problems in the individual schedules with specific resources. You can indicate the changes and see the reaction of the project leaders. You can employ the same approach to see the impact of issue resolution and solutions. There are several goals in carrying out this exercise. First, you are attempting to share resources across the projects so as to improve the schedules. Second, you are trying to get the project leaders to work together more closely in a collaborative mode.

How Should You Communicate Effectively with Management?

This analysis leads to management presentations. Several types of presentations are used: informal presentations of status and issues, formal status reporting, and presentations to obtain resources, change schedules, or resolve issues. Let's consider each of these.

Informal Communication with Management

Meet informally with your manager each week or every other week to give status. This can be done early in the morning in a one-on-one informal setting. A formal meeting is not necessary. What do you discuss?

- Communicate the status of the project(s) in terms of budget and schedule.
- Warn them that there are specific issues that must be addressed at some point in the future. Indicate the impact of the issue on the schedules. However, also point out that you are considering several alternative solutions that will be presented to them.

This informal communication can pave the way for considering the issues later in more formal settings. The approach also gives managers a comfort level about the project. They might suggest some actions or remark on some possible changes. All of this goes toward getting them to participate in the project at a management level.

Formal Communication about Project Status

Systems and technology projects are a challenge for management to understand and grapple with. Many managers lack a technical background. Even if one manager has a technical background, others will not. How do you convey information on technical projects to nontechnical people? From experience, the use of a standardized one-page form for reporting on each individual project, regardless of size, is a good approach. This form is shown in Figure 8.3 for a sample project. The areas of the form are as follows:

- *Title, purpose, scope, leader, and date.* It is important to always reinforce the purpose and scope of a project because the elapsed time of the project may be long. Management must be kept aware of any changes in requirements to the project.
- *Budget versus actual.* This is a chart that plots the cumulative actual costs of the project as well as planned costs. An example is given in the form.
- *Summary GANTT chart.* This highlights the baseline and actual schedule on the same chart.
- *Key active issues.* This area lists the major open issues that remain unresolved along with their impact if not resolved.
- *Major accomplishments.* This is a list of the major achievements and milestones since the last report.

Title: _____ Leader: _____ Date: _____

Purpose: _____

Scope: _____

Summary GANTT Chart   Cumulative Budget versus Actual

A GANTT chart with summary
tasks; the baseline and
the actual appears here.   Money

                                     Time

Active Issues
Issue: Impact:

_____   _____

_____   _____

Accomplishments: _____

_____

Upcoming Milestones: _____

_____

Figure 8.3   Project Reporting Form

- *Upcoming milestones.* This lists what achievements and milestones you
  anticipate in the period.

At a glance managers can see what is going on with the project. Their eyes
will be drawn to the quantitative data first (budget versus actual and GANTT).
Then they will see the issues. This will make them more aware of the issues to be
handled. They will also see the progress that is being made.

What do you do with multiple projects? Well, you already have provided the
individual project information. For status you want to show the benefits and impact of the projects as being greater than the sum of the parts.

If the projects have similar structure, then you can develop the following table. This shows the projects listed as rows. The columns are the high-level milestones. The table shows the estimated completion date or code for completion.

Projects	Analysis	Design
Client-server	C	March 31
Data warehouse	February 15	May 5

This is an easy way for management to track milestones.

You can also show common issues by using the following table. The rows are issues and the columns are projects. This table highlights the impact of the issue on the project. A dashed line or blank can be used if the issue has no effect on the project.

	Projects	
Issue projects	Project 1	Project 2
Issue 1		
Issue 2		

This table shows how the projects link together through the same issues. It will help you in gaining management support for resolving the issue.

A third table shows how projects share common resources. The rows are resources and the columns are projects. The table entry is the quantity and the type of resource being used by the projects. This table helps to show how you are trying to share resources across projects for synergy.

	Projects	
Resources projects	Project 1	Project 2
Resource 1		
Resource 2		

Presentations on Issues and Plan Changes

You have analyzed the issues and the schedule. You have information ready to go to management. How do you present it in a manner that will yield support for your proposed actions? Here is an outline of an approach that works:

- Summary of status of project(s)
- Issue(s) to be addressed
- Impact on the project(s) and business unit if the issue is not addressed
- Benefit to the project, business unit, and organization if the issue is resolved
- Alternative approaches for resolution
- Recommended decision, follow-up actions, and timing

If this looks like the outline of a television commercial, it is not by chance. Note that nowhere is the technical side discussed. You first get to the issue and its impact. For instance, in a commercial for aspirin, the issue would be someone with a headache. Next, show the benefits of having a solution. The ad shows how the person feels and looks better with the treatment. You want to show that you did your homework and considered several alternatives.

Some guidelines are appropriate here. When discussing the impact of the issue, focus on the business side of the impact in terms of delays and effect on business costs. If you consider only the project, the effect will be diminished. The same applies to benefits. With respect to alternatives, consider the following:

- Do nothing. That is, take no action that requires money or slippage. What happens to the schedule? This is the case of restructuring the schedule.
- Throw money at the project. This is the alternative of applying nonpersonnel resources to the project.
- Apply additional resources to the project.

The decision you recommend will need follow-up actions. These might include reassignment of staff, procuring hardware or software, or getting a consultant or contractor. Have these actions ready to go. You want to be able to act on the decision as soon as it has been approved. You don't want to have to come back to management for approval. Keep in mind that there will be a tendency to delay a decision that requires more resources. Managers may not want to take the risk to make a decision. This is countered by the impact of the issue if not resolved.

For multiple projects, you may wish to focus on common issues and resources. Here are some examples:

- A combined GANTT chart that shows the use of a specific resource across several projects.
- A combined GANTT chart that shows the impact as well as resolution of an issue on the projects.

What do you do if the project is now to be changed due to requirements, resources, or some other issue? After the presentation to and approval by management, you are now ready to implement the actions and inform the project team or the project leaders. Here are some steps to take:

- Prepare all paperwork related to acquisition of resources immediately, and get it started through the system.
- Prepare a new schedule with the new dates, milestones, tasks, and so on.
- Update the list of issues with what has been decided.

You are now ready to present your plan to the team or group of project leaders. A lesson learned here is to review it first with each manager or team member one on one. This will allow them to ask questions and punch holes in what you have done. Then you can present it to the group. Here is an outline of your presentation:

- Summarize the issues and what was being addressed.
- Highlight the impacts again of not getting a solution.
- Present the new schedule, which gives the audience an overall view of the situation.
- Define the specific actions that will be taken and discuss how they impact the audience as well as the schedule.
- Indicate how the actions will be followed up in terms of measurement and actions.
- Just passing out a new schedule does not cut it. You must indicate how you will determine if the actions worked.

How Do You Cope with a Project Crisis?

The preceding discussion will work for most situations. However, in rare occasions there is a true crisis. Action is required quickly, and there is little time for analysis. A crisis typically occurs because either a new issue arose or an existing issue reached critical mass. Many people have a tendency to panic and jump at a solution that involves throwing money at the problem. Don't fall into this trap.

If you encounter a crisis, think about a series of changes that can be made to resources, the schedule, personnel, system requirements, business requirements, the business unit role, and so on. Don't restrict yourself. This is your chance to get a number of things in the project and across multiple projects fixed at one time. Use the crisis as an opportunity. Dealing with the crisis means going through the same steps that were discussed earlier.

How Can You Effectively Employ Lessons Learned?

"Lessons learned" is a popular phrase. As mentioned earlier, it is now more formally part of "knowledge management." Whatever its name, there are several challenges in using lessons learned effectively. How will lessons learned be identified? How will they be retained? How will they be applied to the project? What will be the follow-up to see if they did any good? The first two questions were addressed in previous chapters. Here it is assumed that there is a body of lessons learned. To apply these lessons to a project, you must apply them to specific tasks. This is not as easy as it sounds. Start by associating each lesson learned with tasks in the project template and then in the plan. With these lessons learned, you are now in a position to apply the information. Get the analysts and programmers together and discuss each lesson learned that applies to the task. Have them as a group indicate how they might apply the information and what they expect in terms of benefit. Some people may be resistant to changing their methods. They are free to propose alternatives and argue for them. Either way, the methods and tools used in the project are discussed openly between junior and senior team members. This serves to educate the junior members. It also leads to measuring the benefit and effect of the lesson learned later.

EXAMPLE: BEAUMONT INSURANCE

Beaumont Insurance had two groupings of development and enhancement projects. Rather than manage each project separately, it was decided to group them. A group leader was appointed for each. The responsibilities of the group leader included the following:

- Identifying and addressing issues across multiple projects
- Gathering and sharing lessons learned among the projects
- Sharing resources among projects

The project leaders were still held accountable for the project milestones given their resources. At first there was resistance from both project leaders and team members. Project leaders felt that their roles were diminished. The group project or superproject leader continually had to reinforce the role of the project leader. It took time to establish a habitual pattern of behavior.

The team members felt that this was an unnecessary layer of management. A useful response was to show how the superproject leader can gain access to management and to resources more readily than could individual project leaders.

What were the benefits of grouping the projects? There was more sharing of resources in the areas of testing, network support, analysis, and data management across the projects. In the past, some projects finished early and others late. With the superproject approach, projects tended to finish on time more often. There were also fewer surprises for management in terms of issues. It was found that about 20% of the issues were shared by at least three projects. The fear of additional overhead vanished when it became obvious that the superproject leader performed different roles from the project leaders. Another benefit was that the business unit could contact one manager to address a specific problem. Technology experience and selection was more open across multiple projects. This was particularly valuable in sharing client-server and intranet experience.

E-BUSINESS LESSONS LEARNED

E-business project tracking can be very demanding. Following are some specific comments.

- Managers will ask for many presentations since they view e-business as important. You must be ready with organized and canned presentations at all times. You do not have time to customize each presentation.
- The issues tracking and graphs discussed earlier are valuable in e-business since they provide management and business units as well as the team with an assessment of what is going on and where attention is needed.
- Using a common issues database allows you to determine the effect of single and multiple issues across the e-business subprojects.

GUIDELINES

- At the start of the project, take some time to explain to the team members how you will be managing issues, reporting status, and working with them. Many people assume that team members automatically know what will happen. This is not true because this is usually the first time that this particular group of people have worked together.
- Eliminate any project meetings that involve status. Devote project meetings to technical issues, lessons learned, and how methods and tools will be used in the project. This will reduce the overhead and perceived overhead of project management on the team. It will also reinforce the team's view that you can care about results and not the project management process.
- Actively pursue informal communications with several managers. This includes several business unit managers as well as the managers in information systems. This will broaden the base of support for the project.
- If you are managing multiple projects, manage them at several levels. On the general level, focus on the issues that cross the projects. At the detailed level, plunge into a project with the project leader and get involved in the detail. Do this on a consistent basis so that the team will not think that you are pushing aside the project leader. Indicate to the project leader that you want to use a hands-on management style. Attend project meetings unannounced.
- Foster an atmosphere of applying lessons learned across different projects. This can be accomplished at several levels. One is the technical sharing of information among programmers or analysts. Another is between project leaders. Consider having regular lessons learned sessions among project leaders.
- In management presentations on issues have managers or senior staff from the business unit there to emphasize the impact on the business unit if the issue is not addressed, as well as to support early resolution.

WHAT TO DO NEXT

1. Review how your current projects are reported to management. Address the following questions:
   - How are issues associated with projects and schedules?
   - What analysis is performed across multiple projects?
   - How are resources allocated across multiple projects?
2. What process is followed for reviewing milestones? When is a milestone considered sufficiently important for measurement of quality? What constitutes acceptable quality? How are the results of milestone reviews disseminated and followed up on?

SUMMARY

Managing an ongoing project is much more than doing work on the project and calculating and reporting status. In a systems project, you must be proactive by looking to the future and identifying and addressing issues that raise the risk of the associated tasks. Don't wait until an issue surfaces on its own. Work of the project leader includes constant efforts to build the "team" as a team. This is especially important in systems and technology projects where much of the work is perceived as an individual effort. You must build the team so that you raise the level of the junior team members and reinforce the value and self-esteem of the senior team members.

Chapter 9

Software Development

INTRODUCTION

MODERN VERSUS TRADITIONAL DEVELOPMENT

Many system development failures have been reported. Client-server project failures have been reported as high as 50% or more. Data warehousing failures have been cited at similar levels. Traditional mainframe development projects have also suffered problems. Why is the failure rate so high? Although each situation is unique, it seems that many firms are still using the development cycle of the 1960s in the development projects of the 1990s and 21st century.

What are some of the differences in development approaches?

- The traditional life cycle follows a sequential set of phases. Requirements, for example, are assumed to be nailed down prior to design. When requirements later change, the design and programming may have to be redone or reworked, causing the project schedules to slip. In intranet and client-server systems, you learn what a business unit wants by seeing its reaction to prototypes. Requirements remain more flexible and fluid.
- Using object-oriented methods, each transaction falls within an object class. This is almost the opposite of the traditional large COBOL program with numerous IF . . . THEN . . . ELSE statements. Design and programming are more modular in modern systems so that when you discover an exception transaction, you can add the corresponding program objects and carry out limited testing. On the other hand, the COBOL program will have to be enhanced, potentially tearing it apart and necessitating extensive testing.

In previous decades a new system was created that possessed few links to other systems. What links there were consisted mostly of batch processing. The situation today is different. You often cannot afford to replace all legacy systems. Information is more integrated. You are forced to integrate the new system with several legacy systems. The situation often is even more complex because of multiple hardware platforms linked through a heterogeneous network. These factors push us to consider newer development methods and approaches.

METHODS AND TOOLS

Over the years many methods and tools for development have emerged. Tools are software packages that aid in the performance of specific tasks. Methods are approaches to undertaking design, development, and other work. Tools support methods. Failure can occur because of some of the following factors:

- The method is unsupported by a tool so that the staff must resort to manual methods.
- The tools are not compatible with each other, resulting in substantial duplication and even contradictory work.
- The method or tool is inflexible, making it inappropriate for some projects (for example, large or small projects).
- There is no or inadequate vendor support for the tool.
- The learning curve and time consumed using the tool outweigh the benefits.

The more successful tools have tended to aid programming (editors, compilers, interpreters, debuggers, testers, documentation, and so on), system management (such as network management and configuration management), and reduced programming (such as object-oriented tools, fourth-generation languages [4GLs], and database management systems). Tools to aid system analysis and design (for example, CASE (computer-aided software engineering] methods, diagramming, and documenting) have been less successful. Tools that generate code from design input have been developed, but these have met with limited success. This is because the resulting code must be customized and completed.

Even with various methods and tools, system development efforts and projects have suffered from the following complaints and concerns:

- Business requirements and, hence, system requirements change.
- There are sometimes missing methods, tools, or gaps between them that necessitate manual, ad hoc work.
- Benefits from the system resulting from the project are less than estimated.
- Schedules stretch out.

- The project scope expands and changes.
- The role of the business unit and its staff is vague and ill defined.

There is no magic solution to addressing these issues in the form of off-the-shelf tools. Therefore, you must address these factors through project management of the overall development process.

APPROACH

REQUIREMENTS FOR A DEVELOPMENT APPROACH

Let's begin by considering what development will be based on. Traditionally, you go out and collect requirements. You obtain many of these from the business unit staff assigned to the project. However, requirements, if just stated by business staff and not verified, are subject to being inaccurate and incomplete. Remember that there is no penalty for giving partial or incomplete information. Moreover, requirements are often not verified until the entire set of requirements is presented to the business unit for approval and sign-off.

Where will you determine and evaluate requirements? Requirements reflect the activities of the business. As such, it makes sense to observe and analyze the current business process. Requirements for both an improved process and system can be generated. It is also evident that when you observe a business process, you are seeing discrete work—in systems terms transactions. Include management information, reporting, and other work using the systems as transactions. Batch processing is embedded as steps in update transactions (labeled as batch). If there were no current business process, then you would invent the process first to define requirements for a supporting system.

Many types of transactions are available for use, including both regular and exception transactions. In order to gather them all, you have to spend effort over an extensive elapsed time. Thus, if you take a totally sequential approach, your project will likely fail because the time will run out before all of the requirements are known.

Now move ahead to design, development, integration, and testing. If these are treated sequentially, the project elapsed time extends even further. Your approach must be modular to take advantage of the transaction information as it is developed. Information on separate, unrelated transactions gathered later should not disrupt what you have already done. The approach must also be parallel so that additional analysis, design, development, and integration can proceed to the extent possible at the same time. You also have to be open to new methods and tools that are becoming available for intranets, client-server, and object-oriented development.

Let's summarize these factors that impact development:

- Base requirements on information about the business process gathered directly or verified with the process.
- Support concurrent design and development for transactions that have already been defined along with continued data collection for other transactions.
- Accommodate linkage between the new and current (including legacy) systems in design efforts early.
- Relate benefits of the system investment to the combined improved business process and the new system.
- Provide an effective management and control approach for development.
- Given that the transactions in the new process will closely integrate manual and automated steps, it is logical that the user procedures be included in the operations manual with the process.

STEPS IN A DEVELOPMENT APPROACH

The method described in the following steps is targeted at the preceding factors. These steps are not carried out in sequential order. Experience has shown that work can be going on concurrently in several different steps at the same time.

- *Step 1:* Understand the current business process and transactions as well as current systems in terms of potential interfaces.
- *Step 2:* Define a new or modified transaction and then business process.
- *Step 3:* Determine the benefits, user requirements, and system requirements from steps 1 and 2 for both new development and modifications and interfaces to current systems.
- *Step 4:* Design the new transactions and system as well as the interfaces with the current systems.
- *Step 5:* Undertake development, testing, and integration.
- *Step 6:* Prepare business operations manuals (which include user procedures), training materials, and operating procedures.
- *Step 7:* Carry out conversion and data setup for the new system and process.
- *Step 8:* Establish business policies, carry out training, and convert to the new process and system.

For each step it is necessary to answer some or all of the following questions:

- What are the end products or deliverable items?
- What are the roles, responsibilities, and tasks involved?
- What are the criteria for success?

- How should work be performed?
- How should the work be managed?

What are appropriate methods and tools to support the step? Some are established in your organization, so they will not be covered here. Guidelines will be covered separately at the end of the chapter.

What are some of the implications of using these steps?

- The burden of managing the project is more intense than traditional development because the approach is conducted in parallel with multiple steps being active.
- The steps support the modular approach mentioned earlier.
- Requirements changes are less because of the way the work is performed with the business unit staff more involved in the project.
- New methods and tools can be applied in any given step.

STEP 1: UNDERSTAND THE CURRENT BUSINESS PROCESS AND SYSTEMS

End Products

- Operations procedures or manual of the current process including exceptions as well as issues in the current process. This manual documents the current business process in terms of transactions and so provides up-to-date user information. It explains how the current system is used. This can be a useful side benefit to the business unit from the project, which may never have had such documentation before.
- Interfaces involving and using the current systems. This includes documenting the interfaces in place along with identifying issues, problems, and opportunities for improvement.

Roles and Responsibilities

Although a systems analyst or project leader can begin the work and provide guidance, it is essential that the employees of the business unit perform much of the effort. This has several benefits. First, the employees develop a more formal understanding of the work. Second, they understand the problems with the current situation and can relate these problems to the process and system. They are more eager for change. Third, they become more committed and involved in the project. Given the extensive effort required, consider employing junior staff in the business unit who have more available time. Systems staff will have to support the definition of the interfaces.

Criteria for Success

One criterion is the rapid collection and documentation of the work. A second is increasing enthusiasm among business unit employees regarding the project. In addition, you flesh out more exception transactions that lead to having more complete requirements.

Work Methods

Consider building the table shown in Figure 9.1 for each transaction. In this table a transaction is divided into discrete steps that can be manual or automated. For each step, identify who (person, computer, customer, and so on) performs the transaction and what actions are performed. These two columns are based on the documentation method known as Playscript. You can supply calculation and logic rules in the business rules column. The current automation for the step is described in a column. The column of issues identifies both problems and opportunities. The last column is for comments. You might want to include an additional column on the degree of risk or uncertainty in the definition of the transaction. You must admit if you are not certain about specific issues so that you can concentrate on these areas in this and the next step.

Completing this table requires that you carefully cover each significant action. Include such actions as opening mail and contacting customers, but do not get down to hand movements (as in the old days of industrial engineering). From lessons learned it is suggested that you define an initial list of transaction types. Then build the table for several transactions with business staff and conduct a review. These transactions can serve as a model for other transactions, thereby reducing the review effort and improving quality.

Note that you can generate the basis for operating procedures and the operating manual using columns 1, 2, 3, 5, and 6. This is in line with the goal of trying to derive as many benefits from the work as possible.

Transaction: _____

Step	Who	What	Business Rules	Current Automation	Issues	Comments

Figure 9.1   Existing Transactions

Management of Work

The analyst or project leader establishes the first transaction of each type as well as the overall list of transactions working together with the business unit staff. Coordinating the work of the business unit staff in building additional transactions is accompanied by directing reviews of the work. Reviews may typically add or combine steps as well as lead to the identification of more transactions. Each major exception has its own table. Small variations can be covered within the business rules column.

Methods and Tools

The most useful method is to develop an initial set of transaction tables with reviews. This will help in establishing work patterns. A spreadsheet file or database management system application on a network server will suffice for a tool. Team members enter their own information. Network access supports reviews.

STEP 2: DEFINE THE NEW OR MODIFIED BUSINESS PROCESS

End Products

- New transactions based on the new system and improved business process. These transactions are developed by considering several alternatives and then documenting the one that is selected. This end product is critical to determining benefits, identifying the difference between the current and new transactions, specifying the automation required in each step, and preparing the new operations manual.
- Information requirements for interfaces between the new and existing systems. These include data elements, edits, timing, backup, frequency, and volume for each transaction type.

Roles and Responsibilities

Early in step 1 the project leader or analyst can define several alternative approaches for defining transactions. Once an approach is defined and is acceptable, then the business unit staff can participate in defining the new transactions. The project leader can track down the details of interfaces with programmers.

Criteria for Success

All new transactions must be defined on a timely basis as well as being complete and correct. Success, though, means that the new process is significantly better than the current process so as to justify the cost of the system and change. If the benefits do not appear to be sufficient now when you have creatively defined the new process, then you may see even fewer benefits later when the constraints of the technology come down on the process.

Work Methods

Construct a table similar to that used in step 1, as in Figure 9.2. The first four columns will feed the new operations manual and user procedures. This allows business rules to be modified. The table supports a linking between the old and new process steps. This column entry can be blank for a new step or can have one or more entries. The system requirements pertain in most cases to software. The benefits column relates to business benefits. In the comments (last column), address how the issues in step 1 are handled in the new work flow.

Use several different ways to address each transaction, including the following:

- *Improvement with no additional automation.* This shows you how far you can push the envelope without implementing the new system.
- *Improvement maximizing automation.* This pushes the boundary of the technology.
- *Improvement based on simplifying the degree of user expertise needed.* This tests how the process can be stripped down to essentials.

Once you have carried this out for a number of transactions, you can select the approach with the greatest overall benefits. Keep the others in reserve if needed later.

The system requirements can be aggregated at the transaction level first. Then you can work with the volume and performance information specified in the first step to determine the requirements for the system and for interfaces.

Management of Work

Whereas the first step required detailed analytical work, this step benefits from creativity in determining how to simplify or automate the transactions. The project leader can lead and conduct a simulation of the alternatives of the new transaction.

Step	Who	What	Business Rules	Old Process	System Requirements	Benefits	Comments

Figure 9.2   New Transactions

After this creative part, you now must push ahead to define as many other transactions as possible. Concentrate on having a range of transactions as well as including those that have the greatest uncertainty or risk, due to a lack of knowledge, for example.

Methods and Tools

The same methods and tools followed in step 1 can be employed here. Again, make sure that all data are based on the network so that people can share information. If you employ a database management system for the table, then you can combine and analyze the two tables more easily. Dialogue among team members can be recorded for later use because it will contain lessons learned.

STEP 3: DETERMINE THE BENEFITS, USER REQUIREMENTS, AND SYSTEM REQUIREMENTS

End Products

- Comparison of current and new processes and requirements
- Estimation of benefits to the business based on the comparison
- User requirements that include process, organization, and personnel as well as those pertaining to automation
- System requirements for new system as well as interfaces
- Update of the estimated costs of the project and system based on the requirements

Roles and Responsibilities

Business unit staff must be very heavily involved in this step. It is here, after all, that the benefits are pinned down along with the new process. Gathering any additional information related to volumes of work, estimated savings, and so on is the responsibility of the business unit. The analyst can prepare an initial version of user and system requirements for review. Note that in the traditional life cycle this step either is not done or is undertaken by information systems.

Criteria for Success

One measure of success is that there are substantial benefits that will outweigh the costs that will be estimated based on the requirements. What if the benefits do not appear to be sufficient? First, determine where the greatest areas of costs lie. Can these be reduced? Second, determine if the new process can be scaled back to require a less costly solution.

Work Methods

Benefits have been identified for each transaction so that it is possible to generate benefits at the transaction level. What must be added are volumes, personnel time, the value of the benefits, and costs. The table in Figure 9.3 can serve as a starting point. This table builds the benefits at the transaction level. The second column summarizes the benefits from step 2. Monetary benefit along with the justification and reasoning are the other two columns. Note that justification and reasoning will be extremely important later when you present the results of the work to management for approval and credibility.

The transactions can now be rolled up in the table in Figure 9.4. However, you must grasp the overall benefit at the function level to determine actual savings. Use Figure 9.5 to make this determination. Function can refer to a position or group within the business unit.

To obtain user and system requirements, proceed in the same way as you did for evaluating the benefits. Consider constructing the table in Figure 9.6 first. Then move up to the function level to generate Figure 9.7. You can then invert the table to obtain the tables in Figure 9.8 and 9.9 for user and system requirements, respectively.

Transaction: _____

Step	Benefit	$ Benefit	Justification/Reason

Figure 9.3   Individual Transaction Benefits

Transaction	Benefit	$ Benefit	Justification/Reason

Figure 9.4   Benefits by Transaction

Function: _____

Transaction	Volume	$ Benefit	Comments

Figure 9.5   Analysis by Function

Transaction	User Requirements	System Requirements	Comments

Figure 9.6   Requirements by Transaction

Function	User Requirements	System Requirements	Comments

Figure 9.7   Requirements by Function

User Requirements	Function	Comments

Figure 9.8   User Requirements

System Requirements	Function	Comments

Figure 9.9   System Requirements

Management of Work

From experience it is recommended that the project leader or analyst work with the business unit staff and managers to develop the tables. Information technology can function in a coordination and support role, while business unit staff define benefits.

A lesson learned here is that after the first time this is done, the benefits will probably not match up to the costs. This has the positive effect of encouraging the business unit to consider changes in organization, policy, and so on to achieve greater benefits.

Methods and Tools

The tables can be based on the network server. Work from earlier steps should be available. As group members create new tables, they should have space to indicate their reasoning and logic in construction. In a conference room, establish a PC connected to the network with a projector so that the group can review the tables together.

STEP 4: DESIGN THE NEW TRANSACTIONS, SYSTEM, AND INTERFACES

The following activities can be performed in parallel assuming that the staff are available:

- Design of interfaces to current systems
- Design of the graphic user interface (GUI)
- Design of a representative transaction within classes of transactions
- Design of the database
- Design of batch programs
- Design a prototype of the system
- Pilot of the process and system

The second task provides the standard to simplify and make uniform the user interface. The third task provides the start of an object-oriented approach. Add more activities for client-server and intranet applications. For both, tools will have to be acquired and installed. Staff have to be trained. In client-server applications the middleware will be designed. The prototype tests the design of the system. The pilot tests the combination of the system and the new process.

End Products

The end products of this step include the following:

- Database design
- Interface design for links to current systems
- Design of representative transactions (objects)
- Design of batch programs
- Design of new interfaces
- Prototype results
- Pilot results
- Mapping of the design against the requirements and process
- Update of benefits and costs

Roles and Responsibilities

On the surface, much of the review is in the hands of the systems organization. What can the business unit do? Well, remember that steps 1, 2, and 3 are not dead. Work must continue on the transactions not defined at the start of this step. In the prototype and pilot areas, there is much for the business unit staff members to do. First, they can test the prototype system as part of the team. Remember that this is restricted to the system. Then they can fully participate in the pilot because it tests the prototype system in the context of the new process. The pilot helps to validate the system and process together.

Criteria for Success

Success in design can be reflected in quality end products and in integration. You can really add that success is when the benefits and costs fall in line or are better than original estimates. Success is also the acceptance of the prototype and pilot.

Work Methods and Methods and Tools

Follow the accepted methods of design in your organization. In terms of the prototype and pilot, here are some guidelines.

- The prototype includes complete transactions except for the interface to the existing systems. In that way, you can simulate the entire transaction.
- The pilot includes all manual and automated steps for the transactions covered in the prototype.
- The pilot is tested with many employees in the business unit. This not only validates the system and process, but it also acts to build grassroots support for the new process and for change.

Management of Work

The key role is that of the project leader. The project leader must not only co-ordinate work in this step, but also the ongoing work in the previous three steps. Business unit managers are to be involved in conducting and supporting the pilot effort.

STEP 5: DEVELOP, INTEGRATE, AND TEST THE SYSTEM

Here are the end products:

- Completed application system
- Completed system interfaces

The techniques used depend on your environment. In an object-oriented approach, keys to successful development are configuration management and control. The initial transaction in each class can be tested and evaluated. As information becomes available on other transactions, then these transactions can be programmed as objects within their respective groups.

Successful project management means that you will have to oversee the testing and configuration controls. More standard supervision applies to the interfaces and to batch programs. Base your transaction testing on test scripts generated by the business unit. These rely on work in step 2 and help to validate the results in the earlier steps.

STEP 6: PREPARE OPERATIONS MANUALS, TRAINING MATERIALS, AND OPERATIONS PROCEDURES

The operations manual and training materials can be based on the work in step 2 as well as the prototype and pilot of the design step. As the user interface and the exact handling of transactions are known, then this detail can be added. Operations procedures for the system follow internal practices and policies.

It is important in managing this step that you have business unit staff available to generate transaction procedures early, before programming is completed. It is recommended that test scripts, training materials, and operations procedures all be undertaken together as design and development progress. The additional details of how the graphical user interface (GUI) works can be added when available later. In addition to procedures, an effort will be required to define new policies and even job descriptions.

STEP 7: CONVERT AND SET UP THE SYSTEM AND PROCESS FOR OPERATION

Much of the information in the system that is being replaced may be unusable for several reasons:

- Data fields in the new system are incompatible with the current system. Making these compatible can be a major issue.
- Data in the current system may be incomplete or inaccurate. Making manual corrections after conversion is a problem.
- The new system typically has new and different fields than the current system. The content of these fields has to be added.
- The structure of the information may be different in the new than the old.

What do you, as a project leader, do during the conversion step? Early in the project, in the first three steps, evaluate the quality of the data and its completeness in the current system. If it appears that conversion of the data directly is not useful or will only be partial, then consider establishing an initial database for the new system. Include the following tasks:

- Developing software to aid in data cleanup, data field addition, and data entry.
- Working with the business unit to schedule when and what information will be compiled for data entry.
- Obtaining business unit resources for data collection, data entry, and validation.

STEP 8: CONVERT TO THE NEW PROCESS AND SYSTEM

Here the new process and system are turned on. Business unit staff must be trained. All work in previous steps must be completed. The system must pass operational tests. Training must include both the process and the system. The business unit is responsible for training in the business process. Because new employees will be hired after the system is operational, the business unit can also carry out training procedures on the system for its staff. Establishing a method for training the trainer might be useful here.

EXAMPLE: ARCADIA HEALTH SERVICES

Arcadia Health Services was a medium-sized health maintenance organization that survived with legacy systems for patient records, accounting, inventory, and other functions running on a proprietary operating system. For convenience, let time begin at year zero (0). In year 0 Arcadia saw the need to modernize its systems and establish more flexible software.

The legacy systems suffered from the following problems:

- The systems lacked some of the source code.
- It took an average of more than four months to handle a maintenance request because the systems were so old, the staff was new to the systems, and documentation was almost totally lacking.

The systems group reviewed several software packages and made purchasing recommendations to management. Management rejected these recommendations because most funds were committed to expansion. The systems group was not informed of this reason, but instead was told that funds were not available—period. The systems group was told to come back with a proposal for internal development that would cost less money. This revision was approved.

In year 1 analysts collected requirements based on current problems, the current system, and assuming a moderate growth rate. No work flow was analyzed. No assessment was made of the business direction. The requirements and the underlying assumptions never made it to upper management. They were derailed by a middle-level manager who feared that if the requirements were adjusted to fit reality, the project would be dropped.

Senior management meanwhile had decided that Arcadia was at a cross-roads—either it would grow quickly or it would eventually be acquired. The health maintenance industry was undergoing a shakeout. Management chose growth. In year 0, Arcadia had 250,000 customers. This was the volume used by the systems group in estimating volume for the new system. By year 4 Arcadia had 1.5 million customers—an increase of over 400% in fewer than four years. As each new acquisition was made, its customers had to be added to the existing system. This robbed programming resources from development. In desperation the systems manager chose to isolate the development team so as to keep it off limits from the conversion work. When the developers asked how many acquisitions there might be, the systems manager replied that he did not know, but the team had to continue development to meet the schedule.

At the end of year 1, the design was finalized to match the requirements in size with some growth. The new system was based on a 4GL (fourth-generation language) and was to offer many new features and data elements that were not available in the current system. However, the internal staff had to be trained from scratch to learn it, resulting in more delays. Requests for consultants to help in the programming or even to do reviews were turned down.

The new system went on-line in year 4 (more than two years late). It was an immediate disaster. Here are some of the critical technical and management problems that surfaced:

- Management decided to convert all customers in one shot. This was a big mistake because of data problems in the old system. At the time of conversion there were 1.9 million customers spread across more than 45,000 accounts. The conversion program halted whenever it encountered one error in a record. This halted the entire account. After this was fixed, the program was rerun and the next error was hit. Conversion fell behind. It was a nightmare. If an account had 100 errors, the system would halt each time it found one and then processing would have to be repeated after the fix to one record was made.
- There was no cleanup of existing data. The old system tolerated many errors and the new system was unforgiving. For example, in the middle of the conversion it was discovered that the key customer identifier had erroneously been placed into the date of treatment for more than 100,000 customers.
- It took 15 minutes to enter a new customer's data versus a target of 6 seconds.

What was the business fallout?

- Uncollected bills soared to more than $400 million.
- Monies due to care providers (doctors, clinics, etc.) totaled more than $650 million.
- The price of Arcadia stock plummeted by over 50%.

The development project was a disaster. What are some lessons learned from this example?

- Don't freeze requirements—ever.
- Keep communications open between upper management and the development team.
- Make the design flexible for growth.
- If you are going to use new technology, hire someone who knows it.
- Don't isolate the development team.
- Carefully analyze the business process, business rules, and data.
- Plan the conversion. Start with small accounts and build up to larger ones.

E-BUSINESS LESSONS LEARNED

E-business implementation tends to involve buying software tools, using existing systems and interfacing to these systems, and doing some development. Some of the risk management concerns in development are as follows:

- The development effort must be kept limited since there is tremendous time pressure to implement.
- Any development effort must be traded off against integration, testing, and quality assurance work. The more that you develop, the more time you will need for quality assurance.
- The competitive edge in technology for a firm will be in quality assurance and integration, not in the developed software. This is because transactions must be completed in near real time across multiple systems.

GUIDELINES

- Business unit staff members may be reluctant to assume some responsibilities, especially those in steps 1, 2, 7, and 8. There is really no alternative since their participation is essential to gain their commitment. Their commitment is key to the success of the project. Not insisting on full-time participation by the business unit senior staff combined with flexibility in work will help.
- An effective method for carrying out steps 1 and 2 is to divide the team into subteams that review the work of the other groups. From experience this had been found to be more effective than a separate review. It also corrects early any trend toward incompleteness or errors.
- In the initial period there is always a tendency to zero in on the first good alternative to the current process. Avoid this. You may need to change the

workflow during later steps and have some flexibility by having alternatives available.

- Evaluate the new transactions by considering the following:
  — How stable is the new process? How can it evolve without controls if people take short cuts?
  — How will you conduct an audit of the new process?
  — If someone were to create a new exception manually, how could you detect this?
- At the most detailed level in your project plan, put each transaction in the plan as a task. This is important for several reasons. First, it helps you track the work. Second, it shows how important the tasks and transactions are. This is the same approach recommended for programs. Each program, no matter how small, can be a task in the plan.
- Consider establishing a lessons learned database with input from the team members as they work on the steps. This will be a valuable addition later.
- When presenting the new process to management, have the business unit staff members take the lead. After all, they did a lot of the work and they will be very credible. Give specific transactions as examples. Get down to details before you start summarizing the benefits and costs.

WHAT TO DO NEXT

1. Prepare a project plan template for development following the steps presented in this chapter.
2. Identify potential issues and resistance to business unit staff in the expanded business unit role. Discuss how these can be addressed by stressing the positive aspects of involvement.

SUMMARY

To stabilize requirements and ensure benefits, development must involve the business unit staff at all steps. Any sustained absence will create problems later in the project. Changes in the business process are made possible by the new system and modified policies and procedures. Reengineering changes are much more sweeping than this; however, the approach presented in this chapter can be employed within process improvement projects.

Chapter 10

Operations, Maintenance, and Enhancement

INTRODUCTION

Let's begin with some definitions. *Operations* here includes all production support activities. Examples are establishing network support, setting up parameters for a production run, doing recovery and restart, fixing problems encountered during a production run of a system (emergency fixes), doing on-line diagnostics, performance analysis and tuning, and similar work. *Maintenance* is the repair of problems and doing work so that the agreed upon requirements for performance, function, and features are met. Examples of maintenance are repairing errors or bugs in the code, increasing performance of the system to the set performance levels, and even improving user procedures for greater ease of use. *Enhancements* are changes to the system to address new requirements. Handling new features, interfaces, integration, and supporting additional users are examples of enhancements.

Much of systems work is devoted to these activities. In many organizations these activities consume over half of the programming staff time and most of the network and operations support as well. As a system ages, its complexity increases. There is staff turnover, so that over a period of years many different programmers may work on a system. The size of the programs typically grows along with the size of the databases and files. The quality of documentation sinks because little priority is given to its maintenance. It is not surprising that it takes longer and longer to make changes to a system under these conditions, especially when you consider the specific tasks that are performed. In an emergency fix for operations support, a programmer must work under intense pressure to find the source of the error and fix the code. Often there is not time to examine the large programs, so the programmer ends up fixing the symptom of the problem—the problem may still remain. In maintenance and enhancement, the programmer must first review the code to find the areas to be addressed. Then after repair, the code is unit tested. Often there is insufficient time and resources to fully test the program and system, much less to document the changes.

Does it matter how a particular activity is categorized in terms of operations, maintenance, and enhancement? Yes it does. Operations support comes first. The system must perform or be available. Many organizations then rate maintenance second with enhancements coming up last. This is because you can live without the additional features. Given the resources consumed and importance, you might think that this area would receive a great deal of management attention. Such is not the case. People accept the amount of effort as part of doing business—just like business units suffer from business processes that cry out to be fixed. Here are some of the problems encountered in these areas:

- There is a myth that this area is basically incapable of being planned and controlled. Management just assigns a specific level of effort to the work and leaves it alone. The worse case is when management designates specific programmers to work full time on one system in maintenance.
- The resources consumed in these activities deny resources to new projects or major work on current systems.
- Left on their own lower level systems, supervisors sometimes cut deals with their counterparts in the business units. The department gets to keep the status quo along with a few "goodies" while the programmers work on relatively easy work. Upper management in both information systems and the business unit has little awareness of what has been negotiated. This has been observed in several organizations.
- There is little or no cost justification for the work. Every project must define benefits and costs. Yet somehow maintenance and enhancement are often exempt.
- There is a lack of strategy as to what to do with these systems. It is easier to just patch them up each year.

Management gets frustrated when the current systems cannot handle new requirements. This gives encouragement to those who advocate acquiring software packages. However, even when a package is acquired, the same lack of management control persists and impacts the package through interfaces with current systems.

What are your goals for operations, maintenance, and enhancement work? Here is a reasonable list of concerns, which will be addressed in this chapter:

- Define a strategy for each system covering what work will be done in the future, where the system will end up after a certain time, and what resources will be required.

- Implement project management across all of these activities.
- Aim to maximize tangible benefits from work in operations, maintenance, and enhancement.
- Implement more disciplined management of the support-related work.

APPROACH

A PROACTIVE MANAGEMENT APPROACH

If you start with imposing plans and controls on the work in these areas, you will achieve some tactical results, but you may miss the major benefits. You must take a wider view that answers the following questions:

- What business processes and their related systems can benefit from this effort?
- What are the overall plan and concrete goals for operations and maintenance? Should not one objective be to reduce the effort except for emergency fixes?
- How our organization you move away from the policy of dedicating human resources to specific systems?

Whether generated within information systems or by business units, maintenance and enhancement requests must identify benefits and be structured to lead to projects. An example of a form designed to help do this is shown in Figure 10.1. This form identifies the project and provides for its description, justification, benefits, and costs, if approved for action. It also discusses the fit of the request with elements of plans that will be discussed. The item on resource conflicts indicates whether there are competing demands for the same systems resources.

A series of steps can be defined to support a more proactive approach:

- *Step 1:* Measure the allocation of information systems and related resources to determine where they are being allocated (*information systems measurement*).
- *Step 2:* Develop a *process plan* for each critical or key business process. The scope of this plan includes the supporting systems.
- *Step 3:* Define an overall *strategic systems plan.* This combines the results of the process plans with other architecture, development, and software acquisition projects to yield an overall plan.
- *Step 4:* Based on the process and strategic systems plan, determine an automation plan for each major department (*departmental systems plan*). This makes the strategic systems plan more concrete and reveals how the process will be improved through systems.

Maintenance and Enhancement Request

ID: _____   System: _____   Date: _____

Business Process: _____ Submitted by: _____

Description of Change Wanted: _____

_____

Benefit of Change to Business Process: _____

How Benefits Will Be Measured _____

Required by: _____ Reason: _____

Impact, If Not Undertaken: _____

Fit with the Process Plan: _____

Fit with Department Automation Plan: _____

Fit with Strategic Systems Plan: _____

Linkage to Other Requests: _____

Cost of Work: _____

Cost–Benefit Analysis: _____

_____

Resource Conflicts: _____

Action Taken: _____   Date: _____

Reason: _____

Figure 10.1   Sample Form for Maintenance and Enhancement Requests

- *Step 5:* Set priorities for all systems work including planning items from steps 1 through 3, maintenance, enhancements, and operations support. This will be called the *slate of work.*
- *Step 6:* Follow through on the slate of work and measure the results.

These six steps provide for a proactive approach for gaining control and setting priorities. Each step will be discussed with an emphasis on operations, maintenance, and enhancement. Step 1 reveals the current situation and dilution of resources. Step 2 defines an overall plan for critical processes and it helps the organization to set priorities for systems work. Step 3 is the construction of the enterprise systems strategic plan. This is applied to specific departments in step 4. Steps 1, 2, and 4 have direct relevance to departments and help gain support for more proactive work and reduced maintenance. Step 5 is crucial because it is here that the reallocation occurs. Rather than spread out resources across all systems, this step typically focuses resources to gain results for systems that support critical processes. In step 6 you follow through on the allocation and prevent resources from being robbed by ad hoc requests. What will happen if you shortcut this step? Without measurement, you will have no idea of the current situation and the results of doing work. Without the plans for the business process, enterprise, and departments, how can you proactively get ahead of the work and gain business unit support for changes in priorities? If you do not set priorities (step 4), then each request will be considered on its own as it arises—frittering away resources.

STEP 1: MEASURE INFORMATION SYSTEMS ALLOCATION

Let's consider measurement as it relates to operations, maintenance, and enhancement. You can develop several tables here. The first is the allocation of staff to specific systems support.

	Systems			
Staff	System-1	System-2		

Enter the number of hours that each person spends on a given system. Doing this requires either an estimate by supervisors or tracking of personnel time. Next you might want to determine the number of hours spent in operations support, maintenance, and enhancement. This can be too time consuming to gather and track. Instead, you might consider building the following lists. The first is intended to pin down known operational problems. The second provides for tracking maintenance and enhancement requests. Status can be open, closed, being worked, shelved, and combined with another request.

System: _____

Known problems	Severity/impact	Resources required to fix

System: _____

Maintenance/ enhancement requirements	Description	Status	Benefits	Costs	Schedule	Resources assigned

You can now roll up the numbers to obtain a summary. This gives the total effort spread across systems.

Systems	Operations	Maintenance	Enhancement	New development packages

Aggregate the information by business process and by department. In many cases, only one system supports a business process, but there are cases in which the process may use network tools such as groupware as well as an application system.

Business processes	Operations	Maintenance	Enhancement	New development packages

Departments	Operation	Maintenance	Enhancement	New development packages

You can use these tables to answer the following questions.

- Is the systems staff working on the processes and systems that are most important to the organization?
- Are staff members being dedicated to lower priority work because of their in-depth knowledge of the application system?
- Is the split of resources across departments logical, given the differing levels of importance each has to the organization?

STEP 2: DEVELOP PROCESS PLANS FOR KEY PROCESSES

From the standpoint of the business, there must be an overall plan for the top critical business processes. Called the process plan, it defines the future direction for the process (both manual and systems parts) toward which resources can be applied in the future. A process plan contains the following elements:

- Assessment of the current process in terms of issues, problems, and areas of improvement
- Costs of the process as well as systems support
- The long-term new business process that serves as a goal for both the business unit and information systems
- Benefits accruing if the long-term process is achieved
- Systems and technology required to deliver and support the long-term process

The process plan requires work because you must understand the current process and systems, business trends, and new technology and its potential, and you must be able to define a new process. Therefore, with limited resources it is recommended that you spread out the development of the process plans over several years.

What if there is no process plan? The business unit has no direction for the process. Instead, members of the business unit will react to new technology that they see and put in ad hoc, unstructured requests. These will more easily pass through the business unit into information systems. With a process plan in place, there is something to measure a new technology or request against. Will the new technology or request take the process in the right direction?

How do you develop a process plan? Start by identifying 5 to 10 critical processes. Develop one plan and use it as a model for the others. Pick as the first process one that crosses multiple departments and has systems that have substantial interfaces. This will be the most complex situation. For the process, develop the table in Figure 10.2. This gives the steps in the process, their descriptions, current automation, and issues. Each step is described along with the major business rules in the second column. How the current system supports the process is given

Process: _____

Step	Description/ Business Rules	Current System	Exceptions/ Volumes	Issues/ Comments

Figure 10.2 Analysis of the Current Business Process

Process: _____

Issue/ Opportunity	Priority	Impact if not addressed			Benefits			Comments
		Organi- zation	Business Unit	Information Systems	Organi- zation	Business Unit	Information Systems	

Figure 10.3   Issues and Opportunities at the Process Level

in the third column. Exceptions and volume information are listed in the fourth column. The last column provides for issues and comments.

This action helps you understand what is going on at the step level in the process. Next, move up to the process level and identify issues and opportunities associated with the process. Figure 10.3 provides an example. Issues and opportunities can include system replacement, enhancements, maintenance, and operations problems. The impact of the problem on the organization overall, the business unit, and systems are addressed along with benefits. In the comments column you might refer to the specific steps where the impact or benefit can be felt.

The new process can be defined by the following elements:

- Objectives in the design of the new process
- Business assumptions
- Technology assumptions
- Steps in the new process
- Table of analysis (see Figure 10.4)
- Benefits of the new process for the organization, the business unit, and information systems

A potential outline of the process plan document is as follows. The tables presented earlier are included in the plan.

- Introduction
  Purpose and scope
  How the plan was developed
  How the plan is maintained

Process: _____

Step	Automation Needs	Business Needs	Current Process	Issues/Opportunities Addressed

Figure 10.4   Analysis Table of New Process

- Assessment of the current process
  Summary of the current process
  Issues and opportunities
  Impacts, if not addressed
- Long-term business process
  Summary of new process
  Assumptions—business, technology
  New versus current process
- Benefits of new process
  Organization
  Business unit
  Information systems
- Systems and technology requirements for the new process

Note that there is no detailed project plan because this is a strategy document.

STEP 3: CREATE A STRATEGIC SYSTEMS PLAN

Time and space do not allow a step-by-step approach in developing the strategic systems plan. It is assumed that the plan sets forth the following:

- Issues and opportunities facing information systems
- Objectives of information systems
- Constraints (factors that must be accepted because they cannot be changed)
- Strategies (approaches for getting around the constraints and working toward the objectives)
- Action items and project ideas (specific actions and projects that support the strategies)

The plan can draw on the process plans developed in step 2. Using the plan you can develop a table that assists you in evaluating your current and future work. These are given in Figure 10.5. Each issue or opportunity pertaining to operations, maintenance, and enhancement is given in the first column grouped by business process and system. The second and third columns give the issues, opportunities, and objectives that the item supports. The fourth column gives the strategies, and

Operations, Maintenance, Enhancement	Issues/ Opportunities	Objectives	Strategies	Action Items/ Projects

Figure 10.5 Operations, Maintenance, and Enhancement versus Elements of the Strategic Systems Plan

the fifth column lists the action items and project ideas that pertain to this issue or opportunity. If an enhancement has a blank entry, this means that doing the enhancement does not support the strategic systems plan. You have to question whether it should be done.

STEP 4: CONSTRUCT DEPARTMENTAL SYSTEMS PLANS

Department or division automation plans can be initially generated by projecting down the elements of the strategic systems plan and process plans. Using this as a start, you can now identify specific items of relevance to the department. The same tables can be employed as for the strategic systems plan. In both cases, you are attempting to determine the fit of operations, maintenance, and enhancements to the strategic plans. Note that this fit is part of the request form in Figure 10.1.

STEP 5: SET THE SLATE OF WORK

Setting the slate was discussed earlier at a general level. On a quarterly basis, management can review the existing work and requests for changes and set priorities again. The slate of work consists of all approved projects, requests, and other tasks. If there is no organized approach for setting the slate, requests will be handled one at a time. People who scream the loudest or who have the greatest political clout often get their requests approved first. This leaves few resources for the remaining work. It is no wonder that when information systems reported to finance, finance systems received the highest priority. In many cases, the systems of the people with the greatest political muscle are not the ones that should be given attention. It might be more beneficial to first address other systems that cross multiple divisions within the organization.

Over the years, organizations have grappled with the problem of setting priorities. In addition to the reactive mode already mentioned, one approach has been that of fairness. That is, resources are spread across all systems and business units. Each system receives some attention. However, this strategy yields only minor improvements because there is no concentration of resources.

A better approach is to impose a formal review and approval process that cannot easily be overridden unless there is a true emergency. Under this approach, projects and work are only approved if there is a plan of action accompanied by tangible benefits. It is not enough for business unit and systems staff to estimate intangible benefits. These may not come true. The criteria must be tangible benefits and support for the long-term architecture of the organization. The setting of the slate of work must also be accompanied by active measurement to ensure that the benefits were realized. Where do you achieve tangible benefits? A popular myth has been that the benefits are in systems. This is generally not true. The benefits really lie in the business units. They reside in the improvements to business processes. That is where the benefits can be shown and measured.

Let's now examine the process of setting the slate of work. Identify a person who will gather requests and help employees formulate requests. This person who fulfills this centerpost role will provide the following services:

- Offer support for preparing requests
- Review requests for completeness and accuracy
- Track requests
- Develop several "straw man" allocations of resources for management to review
- Measure the benefits of the changes after they have been made

Unless someone is formally named to this role, it will be difficult to support an allocation process. The fallback will put the systems manager in the middle of one-on-one negotiations with business units.

Requests can be categorized as follows:

- Streamlining and cleanup of production systems to reduce operations support
- Maintenance requests
- Enhancement requests to meet government reporting requirements, user requests for changes, and so on
- Backlog of previous requests that were not approved
- Ongoing development projects and installation of software packages
- New development and software acquisition requests
- Policy and procedure changes
- Time set aside for operations support
- Ongoing maintenance and enhancement work
- Implementation of new technology to improve the systems architecture (such as hardware and network upgrades)

With these categories defined, you can proceed with actions to evaluate the requests.

- **Action 1: Review and place requests into categories.**
  Each request is reviewed and placed in the appropriate category. Some requests will be eliminated because they have little benefit. The need for some of the backlog items may have disappeared. Other requests may be combined due to their similarity.

- **Action 2: Evaluate requests in each category.**
  You have a stack of requests in these categories. How do you evaluate them? Each category is unique. For example, policy and procedure changes do not

require money or staff but may require training and other investments. The time set aside for operations support will depend on the system and even the time of year (for example, year-end closing). As each category is discussed, keep in mind that your objective is to weed out as many requests as possible. There are very few resources and many requests. When you consider only critical programmer resources, you are likely to find even more potential conflicts.

—*Streamline production systems.* Programmers here will go into the code and begin to clean up past production problems that were temporarily fixed. The payoff is fewer hours in the future for support and reduced downtime for the application system. This sounds like a good opportunity. However, no new features are added for the business unit. Also, an old system may be in such bad shape that it cannot reasonably be fixed. You will probably only give one to three requests in this category any priority. The others will be eliminated due to resource conflicts, limited benefits, and the condition of the system.

—*Maintenance requests.* These are requests by the business staff to making relatively minor changes in the system. It is very attractive to spend time here because it appears as low risk with some benefit. However, there is often little real benefit. The system is not really improved, nor are there substantial benefits in the long term. Based on these factors you can probably eliminate half or more of the requests.

—*Enhancement requests.* Some of these requests may be mandated and so cannot be avoided. Other requests can be prioritized based on benefits and extent of effort. You want to avoid giving high priority to requests that involve large-scale effort over an extended time (such as more than one person for over three months).

—*Backlog of requests.* A request is in the backlog because it did not receive a high priority. Your goal here is to eliminate such requests. After all, the business unit was able to get by without the change until now. Some of the requests can probably be combined with requests in other categories.

—*Ongoing development and software package implementation.* The tendency here is to automatically approve these projects because they were approved earlier and they are in process. However, this is a good opportunity to revisit the purpose, scope, budget, and progress of the projects. You will find that it is desirable to restrict some of the projects that appear to be dragging out.

—*New development and software acquisition requests.* This is one of the categories that you want to make a priority. Rank requests here by impact on the business and fit with the long-term architecture that you are trying to build.

—*Policy and procedure changes.* Many people don't include these as project ideas. Yet they are, because it may take a project to implement them. By giving the policy and procedure changes visibility, you will accentuate their importance and benefits. You are likely to retain all of these for the final selection.

—*Time for operations support.* This includes normal support as well as emergency fixes. There is a tendency to use last year's experience to estimate requirements for the next period. This is not a good idea because it perpetuates setting aside a "systems slush fund." It is better to cut this back each year.

—*Ongoing maintenance and enhancement work.* Normally, this category is untouchable. However, you do want to evaluate progress, purpose, scope, and benefits. You may find that some work can stop because it yields fewer benefits than originally estimated.

—*Implementation of new technology.* This category often gets little attention. People don't see the benefits of the technology. The technology projects that can be selected are those that are needed to meet performance goals due to workload growth and those that support new systems. You may end up combining some of these items with projects in new development and software acquisition.

- **Action 3: Develop alternative work slates.**
  To help you in developing alternative work slates, you might want to prepare some tables. Figure 10.6 is a basic table. Here you will list the project ideas from each category as rows. The columns are the individual systems staff. In the table entry you will provide the number of hours required by the request.

  Next, you can use this table and separate out the work for which there is little or no conflict. This is because staff members are not interchangeable and are specialized. You will find that because of this, some programmers will be working on lower-priority work. Reconstruct Figure 10.6 so that the requests that are handled by this sorting are at the top of the table. Draw a heavy line below these and the remaining requests.

  There must be an underlying purpose for each alternative. Here are some that have been used in the past.

  —*Focus on maintenance and enhancement.* Under this alternative, you find that there are substantial business pressures to curb costs. The business

Resources

Project Ideas	Person A	Person B	Person C	Person D

Figure 10.6 Project Ideas versus Resources

units are also under pressure to control costs. You are just trying to maintain. There will be no long-term improvement here, but this alternative is useful because it shows what will happen if you stand still.

— *Focus on long-term improvement.* Here you will minimize effort in maintenance and enhancement as well as in production support. Backlog will get a low priority. The resources will go toward new development, software acquisition, and architecture improvement. This is the most aggressive alternative and shows you what can be done if you maximize effort in these categories.

— *Give emphasis to tangible benefits.* Under this alternative you will give highest priority to the requests that yield the greatest tangible benefit.

— *Focus on a specific business area.* Here you will give priority to requests from a specific business unit to which management gives high importance. Minimize work for other areas.

— *Fit the priority with management goals.* Requests that are favored here will fit with management's goals. This can be revenue production, cost control, or improvement in business practices.

- **Action 4: Determine the slate of work.**
Make sure that you define all of the preceding alternatives. Now consider which work slates are similar. You are going to be lucky if the alternatives of long-term improvement, tangible benefits, and management goals are very similar. This is how you will want to present these to management. Experience shows that you will be more successful in marketing if you can show that a particular slate fits several objectives.

STEP 6: FOLLOW THROUGH ON THE WORK

Each approved request should now have a project plan. The project plans should fit into the templates that have already been discussed. Each request should have a designated project leader. You will also want to apply measurements to the work. There is the obvious measurement of tracking the budget versus actual results and the schedule versus the plan. Beyond this is the achievement of the true benefits that were projected when the request was prepared. Remember that the business unit is responsible for achieving the benefits. Information systems can only provide the systems and support. Whether there are increased sales, improved efficiency, or reduced costs will depend on what the business does after the project work is completed. That is why it is recommended that the changes in the business processes, procedures, and policies be made a part of the project. Employ a standard, formal process for assessing the benefits of the work. Figure 10.7 gives a form that can be used for this purpose.

Request: _____ Date: _____

System: _____

Business Unit: _____ Business Process: _____

Estimated Schedule: _____ Cost: _____

Original Benefits: _____

_____

Actual Completion Date: _____ Actual Cost: _____

Reason for Variance with Plan: _____

Benefits Achieved: _____

How Benefits Were Realized: _____

Lessons Learned: _____

Figure 10.7   Measurement of Benefits of a Completed Request

- Justification or explanation is required for variance in schedule and budget.
- In addition to the benefits achieved, the form requires a discussion of how these benefits were achieved.
- There is also space to capture lessons learned from the work.

If there is no measurement, then people just move on to the next project. The knowledge and experience of the previous work are not captured. You then lose a valuable chance to improve the process of maintenance and enhancement as well as other work.

What are some of the reasonable tasks for tracking maintenance and enhancement work? You can generally divide the work into the following categories: initial investigation and review of the code, design and programming changes, testing and reworking the code, documentation, and further testing. In addition, there may need to be training, data conversion, and other tasks as well. Employ a standard template of these tasks. You have several alternatives for having programmers report their time. They can report on time cards the precise tasks that they worked on along with the hours worked. This is often corrupted by people who create a political picture where investigation is reduced and coding is increased. Another approach is to have the programmers put down the total time per project. Then separately, in a project management system, they can input the status of the work.

Managers often treat maintenance and enhancement as stand-alone activities among their programmers. They don't consider collaboration important because the programmers are working on different systems. This is another lost opportunity. Here is an area where you can make a difference. Encourage the programmers to share experiences in making changes. Have them walk through the changes that they made and share lessons learned. This process offers a number of benefits, including the following:

- Lessons learned to improve the work of younger, less senior programmers
- More cross training of programmers so that there is less specialization over time

Maintenance and enhancement are not exempt from issues and lessons learned. There should be a lessons learned database that includes information about each system as well as programming practices. In terms of issues, the programmers can surface a number of problems and potential improvements that can be worked on later if time were available. This is another opportunity to gain information from their work. It also shows programmers that management values their work.

EXAMPLE: RAPID ENERGY COMPANY

Rapid Energy Company is a large multinational energy company. Some years ago the company experienced substantial growth. Information systems had expanded, implemented software packages, and still maintained many old systems. When times were good, no one really paid attention to allocation. Then the price of oil dropped and times got rough. The information systems group was forced to economize. The choice was either to lay people off or to restructure the work or the organization. There had been no measurement and little planning in place. The first step was to measure what was going on. It was found that almost half of the maintenance and enhancement requests could be halted as they yielded little benefit. In addition, when the business units were surveyed and key business processes studied, a number of major needs surfaced that had very tangible labor-savings benefits. Using the priority approach described in this chapter, work was redirected toward these major enhancements. The resulting savings prevented any major layoffs except by attrition. Management support for information systems was greater than it was in the good old days.

E-BUSINESS LESSONS LEARNED

In e-business you should assume that there will be substantial ongoing change and enhancement—more than that for traditional systems. Why? Here are some reasons:

- Marketing will require more creative promotions and discounts to be supported.
- Changes in both the e-business and traditional processes have to be coordinated and synchronized. This requires more systems work.
- E-business is an evolving area so that competition and marketplace pressures will mean more enhancements.
- If the Web traffic grows substantially, there will have to be major hardware and network upgrades.
- Technology is changing and almost seems to be directed at e-business— meaning more work.

That is why we said earlier that e-business is a program that is ongoing.

GUIDELINES

- Each year develop a combined list of benefits for the work performed by information systems. You can create a table in which the rows are the various projects and work and the columns are areas of benefit (cost savings, performance improvement, revenue increase, and so on). Then sort this table by business unit, business process, and technology area. This will give you a better understanding of what the information systems group is accomplishing.
- Visit each programmer individually to discuss that person's technical work. Have each programmer identify issues and lessons learned, and put these results into databases. Next, gather the programmers together in a meeting and show them how they share the same issues and lessons learned. Build informal collaboration and information sharing.
- To support a project management atmosphere you should have each programmer participate in the development of project templates for maintenance and enhancement. You may have a number of different templates that depend on specific types of enhancements.
- Market the changes in the current method of operation through self-interest. The programmers gain by showing that their work is important and valued. The business unit gains by having its high-benefit requests undertaken. Management gains through visibility of the information and achievement of benefits. The information systems unit gains by implementing project management across almost all work.

WHAT TO DO NEXT

1. Measure what you currently have and what is going on. Identify the projects and work, and place them into categories as was discussed earlier.

Then examine the allocation of effort. If this same allocation continued for several years, would there be any major benefit or change?

2. Using the results of the first item, evaluate the benefits of the work. Evaluate the benefits that were achieved from the previous maintenance and enhancement work.

3. Identify potential projects that can be performed if resources were available. Develop an alternative using data from the past to indicate what may have been possible if resources had been redirected toward this more productive work.

4. Examine how people report their time on production, maintenance, and enhancement. Is there sufficient detail to determine what is going on? Are the programmers sharing any lessons learned? Is there an awareness of outstanding issues with each system?

5. Is there a real effort to identify benefits for each request for maintenance and enhancement? Is there any follow-up to determine if the benefits were achieved?

SUMMARY

Many people believe that maintenance and enhancement fall outside of project management. They feel that imposing project management on such small work and limited tasks is too great a burden. However, maintenance and enhancement tasks can stretch out for months. Without any controls or formal approach, a substantial percentage of available resources can be drained off into work of lower priority. In this mode maintenance and enhancement work is reactive. To make it proactive you must actively seek out opportunities for systems work and put them into the planning process. Doing this will give greater visibility to the work as well as increase awareness that tangible benefits count.

Chapter 11

Software Packages

INTRODUCTION

Many firms have acquired or are in the process of replacing some or all of their legacy systems with standardized, integrated software. Examples include SAP, J.D. Edwards, Peoplesoft, and Baan, among others. This trend will continue for years as companies move to employ software packages rather than customized software solutions. The benefits of software packages include the following:

- Reduced or eliminated development effort
- Ability to control staffing levels in information systems related to programming
- Faster availability of software and features
- Reduced risk because of less development
- Less maintenance and perhaps greater flexibility to support enhancements once the new systems are in place

The risks and costs of using packages are still substantial—if the wrong package is selected, the entire project can fail.

- The company must still struggle to be competitive with other firms in the same industry using the same or similar software packages.
- If the business processes and packages don't fit together, then something has to give—the business suffers.
- Changes to the software will have to be made by the vendor or someone approved by the vendor.

The process of selecting and implementing such packages is quite different and more complex than was the case in earlier times. Some of the critical factors are as follows:

- In the past, a package augmented existing business processes. Today, the business processes must fit closely with the package to achieve substantial benefit. As an example, suppose that your firm would like to ship equipment to good customers without the paperwork. However, if the package requires the paperwork first, you cannot really ship the equipment right away.
- The packages are integrated today, which means that their use involves multiple departments and divisions. Departments that never really had to work together now find that, because of software, their patterns of behavior must change. Project management of the implementation process is much more complex.
- Due to the complexity of the software and the long learning curve, most firms find that they have to hire consultants to support the installation process. The cost of the consultants often far exceeds the cost of the software package.
- Implementation can take months or years as company management defines new or modified business processes. Who will perform the normal work during this period?
- The benefits of the software accrue to the company's general management and to the company as a whole more than to individual department staff. Their workload and complexity may actually grow. It is very difficult to motivate the lower levels of a business unit that the package is in their best interests. The effect of a mandate from management can be dissipated over the years of installation.
- The company must endure risk as resources are applied to the implementation. Flexibility and staff resources that are normally available are now consumed in the implementation of the package. Changes to existing software that will be replaced will likely be turned down. Programmers who supported the current legacy systems to be replaced are devoted 100% to the new software package. Do you think that these people are going to eagerly support the package? Their resistance could create another project management problem.
- Interfaces must be built that link legacy systems—which will not be replaced by the software package—and the software package. This can be complex because the content, meaning, and other characteristics of the information may differ between the package and legacy systems.

Why do failure and problems occur? Here are some reasons:

- The culture of the country and that of the firm do not fit with the software. This is discovered too late and the project fails.
- The business units and corporate management are too inflexible to make changes in business processes that will accommodate the software package.
- The company fails to take advantage of the new features of the software and instead thinks of the software as a straight replacement for the existing software.
- The information systems organization is faced with many challenges in implementing a software package:
  —How to maintain current systems during the transition.
  —How to manage such a large project.
  —How to transition current systems staff to the new technology.
  —What to do with the other software not touched by the software package.

The purpose of this chapter is to determine an approach for managing the selection and implementation of large software packages.

There are some basic rules that you should apply to both software package evaluation and implementation. Give high priority in evaluation not to the features pitched by the vendors but instead to whether your business transactions will flow through their systems. During implementation give attention to not only the software, but on how you will fill the gap between what the package delivers and what you need. In some cases, you may have to invent new shadow systems and workarounds—even for the new package.

APPROACH

STEPS IN IMPLEMENTATION

The project can be divided into steps.

- *Step 1:* Assess your current systems, processes, and technology.
- *Step 2:* Evaluate the software packages and vendor support required.
- *Step 3:* Select and negotiate contracts with vendors.
- *Step 4:* Manage the initial software setup and training and conduct a pilot project.
- *Step 5:* Undertake full-scale implementation.

Before proceeding with each step, please note that each situation is unique. The following suggestions will serve as guidelines and lessons learned—not as inflexible rules.

STEP 1: ASSESS YOUR CURRENT SYSTEMS, PROCESSES, AND TECHNOLOGY

Why go through this step? Why not just plunge in and select the software? Knowing what other problems you will face is one reason. Another reason is to prepare the organization for a change in business processes. Here you must uncover and unearth your shadow systems, workarounds, and exceptions.

The end products of the first step include the following:

- Identification of upgrades or replacements for hardware, system software, and network components and systems required to support the software package.
- Determination of the business processes, business units, and managers that will be involved in the other steps.
- Assessment of the scale of effort, costs, and issues that will have to be faced in implementation.
- Estimation of benefits from implementation of the software package.

This step can be undertaken even if there is a management directive to implement a specific software package. The preceding information can contribute later for generating a realistic perspective of the requirements for implementation. Several activities can be performed in parallel.

- **Action 1: From basic requirements of the software packages being considered, determine the potential future hardware, software, and network environment.**
   Most packages can operate on a variety of machines, operating systems, and database management systems. An example might be a Hewlett-Packard, DEC, or IBM UNIX machine with a 4GL such as Informix or Oracle. If you already have such an environment, then you can determine the incremental additions required. In other cases, the company is faced with acquiring a new platform to support the 4GL or database management system. This is a major step because the interfaces required between the old systems and the new package will cross the network with multiple platforms.
   The network and workstations are another part of the puzzle. Although you may be able to support the package in a minimal form with the current network, you are just asking for trouble. The business unit will change its pattern of work by using the network and PCs even more than it currently does. Existing PCs may have to be upgraded or replaced along with network interface cards, hubs, routers, and other network components.
   How are the costs of upgrades accommodated? One view is to associate them with the software package implementation. Another view is to consider the upgrades part of the normal price of doing business using technology. If the upgrade issue is not faced early, then it will hit with full force when the

Dept.	Evaluation Involvement	Implementation-Pilot	Implementation -Full Scale	Business Process Role

Figure 11.1   Table of Department Involvement

new system is being implemented. The implementation schedule will suffer as resources have to be pulled off to handle the upgrade.

From lessons learned, the upgrade can commence after software selection and negotiation. This will support focus on the software later. Network implementation is considered in more detail in Chapter 12. Here it is recommended that the scope of installation, testing, training, and support be identified to assist in planning during later stages.

- **Action 2: Identify the business departments and processes to be affected by the software package.**
  Let's assume that you are considering payroll and human resources software. You might think that only these departments are involved. You might be wrong. Some of the packages move data entry and capture out to line business units. They have to be included in the project as well. In some cases, you might find it easier to determine which departments are not involved. Consider creating a table of department involvement. An example is shown in Figure 11.1. In the second through fourth columns, the roles of the departments can be listed in terms of evaluation and implementation. The last column indicates the department's role with respect to the business process.

  In contacting departments to collect this information, identify people who will be involved in evaluation and implementation. You might wish to use the table in Figure 11.2 for this. It identifies their roles as well as expertise. With respect to business processes identify both formal and informal processes.

- **Action 3: Analyze business processes and transactions in terms of standardization and potential limitations.**
  This can be the most time-consuming action, but it is limited here to getting an initial assessment of uniqueness and culture. Detailed analysis will come

Phase of Work: _____

Dept.	Person	Telephone/ e-mail/fax	Role	Expertise	Comments

Figure 11.2   Roles of Staff

later. Here you have to determine the degree of flexibility of the processes and organizations to accommodate to the packages.

Concentrate on defining detailed transactions in the current business processes that can be employed in the evaluation of the software packages. Gather both typical and exceptional transactions. Collect some sample data as well. For each transaction, identify the steps, business rules, and exceptions that can occur. Try to pin down the reasons for exceptions. A list of tasks for step 1 is given in Figure 11.3

STEP 2: EVALUATE SOFTWARE PACKAGES AND SUPPORT REQUIREMENTS

Traditionally, you define a list of features for the software that you can then prioritize into those that are absolutely required and those that are desirable. Each package can then be evaluated and scored. You can generate a total score by weighting the various features (multiplied by the score). This is most useful if you can identify some features that are very unique. Then the scoring can eliminate some of the alternative software packages. However, it is often the case today that most of the packages are flexible enough to possess many features.

Because the software integrates with the business process, following only this approach may result in problems. You must also determine the fit between the software package and the business process. If possible, take the transactions gathered in the previous step and apply them to the software packages. This may not be possible, because a massive effort may be required to set up the tables and parameters just to get ready to run the transactions. There are several fallback positions. First, working with the vendor, you can simulate how the transactions will be handled. Ask questions about how the package is to be set up and how each step of the transaction will be addressed. A second approach is to interview current users of the software package. Some of the questions to ask are given in Figure 11.4.

Some additional suggestions are as follows:

- Ensure that you include exception transactions as well as normal ones. It is often the exceptions that contain business rules that cannot be handled by the software.
- Make sure that business unit staff members are involved in the evaluation. If they don't participate, they may feel left out at this stage and then feel that the package is being shoved down their throats later.
- Review and document results as you go. Also, record questions and issues that may impact implementation.

What is success? You might say that it is selecting the most cost-effective package that fits the requirements. That definition is insufficient today. The business

1000 Project Planning
    1100 Purpose
    1200 Scope
    1300 Establish Issues Database
    1400 Establish Lessons Learned Database
    1500 Set Up Project Management Software
    1600 Train Project Team in Project Methods and Tools
    1700 Establish Project Reporting Process
    1800 Establish Issues and Lessons Learned Processes
2000 Initial Data Collection for Software Packages
    2100 Survey Software Packages
    2200 Collect Information on Experience and Use
    2300 Define Technology Requirements for Each Package
    2400 Define Business Process Requirements for Each Package
3000 Estimate Future Network, Hardware, and Software Requirements
    3100 Review Current Workload and Systems
    3200 Estimate Future User Base and Volume of the Software Package
    3300 Define Network Structure and Components for the Future
    3400 Identify Hardware Types and System Software for the Future
    3500 Conduct Sizing Analysis to Determine Quantity by Type of Component
    3600 Itemize Upgrades Required
    3700 Define Migration Path from Current Architecture
    3800 Estimate Costs and Define a General Schedule for Migration
4000 Identify Business Units and Processes to Be Affected
    4100 Identify All Departments with Potential Involvement
    4200 Define the Roles and Responsibilities for Each Department
    4300 Develop Approach for How Departments Will Work Together
    4400 Identify Key Individuals by Phase for the Software Package
    4500 Establish Project Team
5000 Analyze Current Business Processes
    5100 Analyze Central Process to Be Impacted by Software Package
        5110 Create a Transaction List
        5120 Analyze Each Transaction
        5130 Define Issues and Opportunities for Improvement with Current Transactions
        5140 Identify Other Manual and Supporting Systems in Addition to Legacy Systems
        5150 Document and Review Transactions
    5200 Analyze Related Processes That Will Interface with the Central Processes
        5210 Identify Exact Interfaces
        5220 Determine Issues, Characteristics, and Other Aspects of Interfaces
    5300 Analyze Existing Software That Will Interface with Software Package
        5310 Determine Current Interfaces
        5320 Assess Ability to Make Changes to Current Systems
    5400 Determine Format, Condition, and Quality of Data in Current Systems
        5410 Identify All Relevant Data Files
        5420 Analyze Each File and Data Element
        5430 Review History Files
6000 Estimate Costs, Benefits, and Schedule of Implementation
    6100 Develop First Cut at Costs for Technology, Systems, and Implementation
    6200 Develop Initial Schedule and Plan
    6300 Define Benefits from Software Package
    6400 Perform Cost-Benefit Analysis

Figure 11.3   Tasks for Step 1

- Technical
  — What are the general alternatives for hardware, system software, and the network?
  — What suggestions would the vendor make regarding your existing configuration?
  — What software is required beyond the software package?
  — What systems management methods and tools are needed?
  — What network management methods and tools are needed?

- Business
  — What business processes does the company's products support?
  — What countries and organization types are supported by the system?
  — How is the system made flexible for the business process?
  — How are exceptions handled?
  — What productivity and activity statistics are available through the system?

- Financial/General
  — What is the financial condition of the firm?
  — How many employees does it have in development and support?
  — What is the direction of development for the firm?
  — What software is under development?
  — Does the firm have any new products or upcoming versions that would cause you to wait for their release?

- Existing Users
  — What are the greatest strengths of the product?
  — What are the weaknesses of the product?
  — What was the quality of support during installation?
  — What was the quality of support during and after implementation?
  — Did the users receive the level of benefits that they anticipated?
  — What were their biggest surprises?
  — How did they organize the implementation project?

Figure 11.4 Questions Related to Software Packages

unit must not only agree, but also commit to implementation and to change. Otherwise, the installation of the "best" software will end up on the rocks.

Let's define the role of the business unit in this step. In the preceding step the business unit participated in defining and assessing the current business process. Here the business unit must be involved in the evaluation through direct participation. During the evaluation there will be times when the current process does not work with the system. The process will have to change. Get this out on the table. Have the business unit react. Otherwise, work-around efforts might be necessary later to circumvent the software package and preserve the current business process. IT cannot agree to the new process and change. That is the responsibility of the business unit, which will have to identify the necessary changes to its pattern of work. Act as a coordinator and scribe to record the results and changes. There may be implications for policies and organization as well as procedures.

You will likely require consulting support to implement major software packages. The identification, evaluation, and selection of the consultants can make or break the project. Many firms select the software first and then find the consultant. This is a bad idea for several reasons. First, the consultant costs will likely exceed the cost of the package many times. Second, success will depend as much on the consultant as on the package. You can also select a package where the assistance in your area or country is limited or where assistance is very expensive. Consider evaluating several consulting firms when you have narrowed the software packages to two alternatives. Some criteria for evaluating consultants are listed in Figure 11.5.

Narrow the candidate software packages to two. Ranking should not be based on cost but on the benefit that the company will derive (as the benefits far outweigh the cost of the software). When considering cost, focus on the total cost that will include hardware, consultants, and internal staff support. Thus, it is important for the business units to define the benefits for each package. A checklist is given in Figure 11.6. Packages that leave the current process intact will appeal to the business unit but will lack sufficient benefits. Management benefits such as availability of centralized information must be quantified. What if management attempts to impose a choice? Include another package and perform the evaluation. You still must evaluate the package against management's choice. Figure 11.7 contains a sample list of tasks for this step.

- History and track record of firm
- Length of time and number of engagements involving the software package
- Qualifications of staff members identified
- Experience with specific firms in industry
- Range and scope of tasks that firm can address
  — Planning for implementation
  — Analysis of business processes
  — Setup of software package
  — Installation support of software
  — Customization of software
  — Training in the software
  — Measurement of results
- Number of engagements currently
- Ability to support multiple locations in different countries
- Status of consultant from the perspective of the software vendor
- Previous clients and recommendations
- Cost
- Availability of staff
- Guarantee of work

Figure 11.5   Criteria for Evaluation of Consultants

- Costs
  — Software package—basic
  — Software package—additional modules
  — Related system software required
  — Related database management system software required
  — Utility software required
  — Network hardware upgrades and purchases
  — Network software upgrades and purchases
  — Server hardware upgrades and purchases
  — Server software upgrades and purchases
  — Client hardware upgrades and purchases
  — Client software upgrades and purchases
  — Labor costs to support upgrades and purchases
  — Vendor software installation support
  — Vendor software implementation support
  — Consulting support—installation
  — Consulting support—implementation
  — Data conversion effort
  — Pilot system implementation and testing
  — Vendor software maintenance and service
  — Hardware maintenance and service
  — Dedicated staff support for software package implementation
  — Ongoing staff support for software package
- Benefits (note that some overlap)
  — Increased work volume per unit of time
  — Reduced paper and handling per transaction
  — Reduced labor costs in business unit
  — Reduced support costs from information systems
  — Reduced hardware maintenance and software support (replacing legacy systems)
  — Reduced error rate
  — Reduced rework
  — Simpler and easier training
  — Automatic productivity and activity statistics
  — Work-flow tracking and ability to locate any transaction
  — Reduced time in data analysis (through better data quality)
  — Improved availability of system
  — Improved reliability and reduced downtime
  — Faster customer service and response time
  — Flatter, streamlined organization

Figure 11.6   Cost and Benefit Elements

21000 Determine Candidate Software Packages/Consultants and Data Collection
    21100 Collect Detailed Information on Each Package
    21200 Define Feature and Capability List
    21300 Collect Information from Existing Users
    21400 Survey Internet/Literature for Reviews and Experience
    21500 Define Evaluation Criteria for Consultants
    21600 Perform Data Collection for Consultants
    21700 Collect Data from Consultant References
22000 Preliminary Evaluation
    22100 Develop Evaluation Approach and Criteria
    22200 Quantitative Scoring and Evaluation
        22210 Define Weighting Method for Features and Capabilities
        22220 Carry Out Scoring
        22230 Document Evaluation Results
    22400 Transaction Testing
        22410 Identify Transactions for Testing
        22420 Document Anticipated Test Results
        22430 Conduct Transaction Testing
        22440 Evaluate Test Results
        22450 Document Test Results
    22500 Conduct Evaluation of Consultants
        22510 Perform Evaluation of Consultants
        22520 Document Results of Analysis
    22600 Prepare Initial Evaluation of Software Packages
        22610 Software Packages Only
        22620 Consultants Only
        22630 Combined Software Packages and Consultants
        22640 Select Two Finalists
        22650 Document Results of Analysis
        22660 Obtain Management Approval

Figure 11.7   Tasks for Evaluation of Software Packages

STEP 3: SELECT THE PACKAGE AND NEGOTIATE THE CONTRACT

Now you have narrowed the field to two packages. You must select one along with the consulting firm and then negotiate with both the software vendor and the consulting firms. Note that the consultant choice is linked to the package choice. The consultant will eventually disappear, so why have the consulting firm involved here? Because of your dependence on the consultant for success of implementation. The package and its support are tied closely together.

Final Selection

With both finalist packages suitable, how are you going to decide which to choose? Here are some guidelines:

- Any cost differential that is present now may disappear as the implementation progresses. Because much of the cost lies in labor associated with implementation and because these costs are often comparable for different packages, cost may not be the deciding factor.
- A deciding factor may be your firm's willingness to change its business processes and policies. Each package will have its own requirement for business change. Business change in turn relates to the degree of benefit. Thus, if two packages have a similar cost to implement, you might decide in favor of the one that offers the most benefit because management is willing to make changes. On the other hand, you might decide in favor of the one that requires less change, if it is politically acceptable despite the fact that it offers fewer benefits.
- Another criterion for selection is that of risk. Risk in software packages will often lie in implementation. Implementation will depend on the combination of the software vendor, the consultants you select, and your staff. Here the consultant alternatives for each package may count more than the actual software.

Contracting

Figure 11.8 provides a list of contract issues for both the software vendor and the consultant. Among the most critical questions in terms of contracting are the following:

- What is the role of the software vendor in providing initial installation and training support?
- What are the responsibilities of the software vendor in fixing problems with the software?
- What is going to be the responsibility of the vendor, consultant, and you in managing the implementation? What roles will each play?
- How will the consultant be measured and monitored?
- How will problems and issues be resolved between the vendor, consultant, and you?

STEP 4: INSTALL THE SOFTWARE AND CONDUCT A PILOT PROJECT

Begin forming the project team. Consider having three project leaders: one from the vendor, one from the consulting firm, and one from your firm. These leaders are

- Software Package
  - — Functions and features included
  - — Hardware, network, and system software environment required
  - — Installation support provided
  - — Implementation support provided
  - — Warranty
  - — Performance standards
  - — Acceptance criteria and testing
  - — Maintenance and ongoing support
  - — Cost structure and discounts
  - — Upgrade policies
- Consultant
  - — Available staff and their qualifications
  - — Types of services to be provided
  - — Ownership of any customized work
  - — Protection of information regarding company operations
  - — Termination
  - — Level of effort
  - — Work quality evaluation
  - — Replacement of staff
  - — Billing and payment terms

Figure 11.8 Contract Issues

not equals because you are the customer. An alternative is to have four leaders: one from the business unit, one from IT, one from the vendor, and one from the consultant. However, this setup can become unwieldy and difficult to manage. Almost all successful implementations involve multiple leaders so that decisions can be made quickly and parallel subprojects can be undertaken.

The roles of each can be defined as follows:

- *Internal information systems.* General project management, design and programming interfaces, operations procedures, data conversion.
- *Business unit.* New operations procedures, training materials, estimation and responsibility for benefits, testing.
- *Software vendor.* Software installation, training, support for testing, troubleshooting, testing.
- *Consultant.* Implementation support, support through lessons learned and experience, training of systems and business unit staff, testing.

Note that there is overlap because some functions involve multiple organizations. The three project leaders develop an overall summary plan that includes the tasks listed in Figure 11.9.

Next, each manager works with his or her own staff members to identify their tasks in detail. Several joint working sessions can be held to refine the tasks and schedule. Attendees also should include business unit staff as well as employees

31000 Project Management
    31100 Identify Roles and Responsibilities
    31200 Identify Team Leaders
    31300 Develop Task Lists
    31400 Define Initial List of Issues
    31500 Establish Detailed Project Plan
    31600 Obtain Management Approval
32000 Software Installation
    32100 Review Environment for Software Package
    32200 Acquire Additional Hardware, Network, and Software to Support Package
    32300 Install and Test Supporting Hardware, Network, and Software
    32400 Train and Prepare for Software Package Installation
    32500 Install Software Package
    32600 Test Application Software in Network
33000 Training and Setup
    33100 Training for Information Systems and Business Unit Staff
    33200 Definition of New Business Process Transactions
    33300 Definition of Parameters and Table Values for System
    33400 Setup of Parameters and Tables in System
    33500 Preliminary Testing of Parameters
    33600 Training of Systems Staff for Interface Programming
    33700 Design of Interface Programs
    33800 Programming and Testing of Interface Programs
    33900 Documentation of Business Process for Pilot
34000 Pilot Project
    34100 Train Business Unit Staff as Part of Pilot
    34200 Load Data and Set Up for Pilot Project
    34300 Establish Network and System Environment for Pilot
    34400 Define Acceptance and Measurement Approach for Pilot
    34500 Conduct Initial Pilot Test
    34600 Review Results and Refine Pilot
    34700 Refine and Redo Pilot
    34800 Measure Results and Estimate Benefits

Figure 11.9   Tasks for Initial Implementation of Software Package

from the other groups. This collaborative approach not only heads off problems later but also helps toward gaining the employees' commitment and involvement. By working together the team will acquire a common vocabulary.

Software Loading

Loading software used to be easier when it resided on a mainframe computer. Today, most software is client-server or intranet based. This means that there are many components and more people involved in setting up the software. You will probably want to establish the software package on a separate testing client-server network. This will not only ensure the integrity of the current network, but it will also support training, development of system management and operations procedures, and testing. You will also want to involve key systems programmers, network support staff, and application programmers in these tasks to aid in their learning.

What are some of the issues that you may face here? One is that you may suddenly find that the network software or hardware must be upgraded to support the software package. A second issue is the challenge of making technical staff available to work with the software package in addition to their normal work.

The end products of the software loading process are as follows:

- Functional software on a network
- Technical staff trained at a basic level in installation and operations
- Established framework for the systems and network management and operations procedures

The procedures can be finished during the pilot.

Training and Setup

Training requirements for integrated software packages are extensive. This is because exact parameters and table values have been loaded into the system. This requires business unit staff and managers who know the current business process in detail, understand what the new process will look like with the new package, and are empowered to make decisions. This is a tall order, given that there are very few of these people and there is a need to continue to support the current process. The role of the consultant is helpful here because he or she can explain what alternative parameters mean in terms of the business.

How can you track the training and setup? The most complete approach is to list each table that must be specified as a task. You will really know where you are and what progress is being made. However, maintaining such a detailed schedule will take time. Going to this level of detail provides some additional benefits. Building the list of tables and tasks with the consultant and business unit staff will show the business unit staff members what is ahead of them. If they are encouraged to update their own schedules, then they can use the schedule as a checklist. In building the tables for the software package, you will come upon unknown factors that require further analysis. Enter these factors into the issues database to ensure that they are addressed later. Then the team can go on to other tables. It is similar with lessons learned. As you learn about a specific table and what its impact is on the system, you can enter this information into the lessons learned database.

What challenges do you face in the setup? One is getting the right human resources from the business units to participate. Another is maintaining a steady effort over several months of setup. It is difficult to maintain momentum when the people still have to perform some or all of their normal tasks. A third challenge is tracking down unresolved issues and questions. Not resolving some key questions can mean that the pilot cannot test this area of the software package.

The end products of training and setup are as follows:

- Business unit managers and staff have been trained in the setup and use of the software package.
- Parameters and tables of the software package are established.
- The software package is ready for the pilot.
- A start has been made on training materials and procedures.
- Decisions have been made on the detailed changes to the business process that will be tested in the pilot project.

Pilot Project

A pilot project is important for any major software package. First, you validate the parameters of the tables of the system as well as the training of the staff. Second, you can use the pilot to build procedures and training materials. Most important, you can determine what changes in the business process will be necessary to fit with the software (the parameters and tables have only limited flexibility). The pilot then develops the new business process and helps to validate the benefits that were estimated when the package was justified to management.

What are some of the risks in the pilot? One is that the pilot does not exercise enough of the features and capabilities of the software package. If this occurs, then the business process will be in a state of flux during implementation—leading to more problems. This happens when management attempts to rush the pilot project so as to begin implementation earlier. Another risk is that the pilot does not exercise all of the exception transactions and conditions of the current business process. A third risk is that the business unit may attempt to run the pilot with the old business process. This may lead to rejection of the software package.

The end products of the pilot project are as follows:

- Validation of the software package tables and parameters
- Definition and validation of the new business process along with the software package
- Completion of operations procedures and training materials for end users
- Completion of systems and network management procedures as well as computer operations procedures
- Trained core of business unit staff who are confident in the software package

The last item is perhaps most important. It is a big leap from a pilot to full-scale implementation. If the pilot yields many unresolved issues, then consider addressing these issues, followed by carrying out a second, limited pilot. Doubts about the system will have a major dampening effect on its implementation.

STEP 5: UNDERTAKE FULL-SCALE IMPLEMENTATION

A list of tasks for this step is given in Figure 11.10. What areas have risk? Several seem to surface repeatedly.

Conversion

While some reference tables are easily converted, the basic files are a different matter. Will the master file be converted? If so, how will missing data be captured and entered? How will the data in the master file be cleaned up? Some

41000 Business Process Change and Preparation
      41100 Review Results of Pilot
      41200 Prepare and Review Procedures and Policies
      41300 Prepare Training Materials
      41400 Determine Migration Approach for Old Data
      41500 Design and Program Conversion Programs
      41600 Establish Data Entry and Cleanup Squad
      41700 Verify That Data Is Established
      41800 Dry Run Test the New Business Process
42000 Network, Hardware, and System Software
      42100 Determine Exact Requirements Based on Pilot
      42200 Procure and Install Hardware, Network, and Support Software
      42300 Conduct Network and Performance Tests
43000 Interface Programs
      43100 Complete All Interface Programs
      43200 Conduct Tests of On-Line and Batch Interfaces
      43300 Conduct Tests for Backup, Restart, and Recovery of Interfaces
44000 Software Package Customization
      44100 Define Customization Requirements Based on Pilot
      44200 Carry Out Design of Programs
      44300 Program and Test Customized Software
      44400 Document and Establish Software for Production
45000 Operations Procedures and Setup
      45100 Define Operations Procedures for Normal Processing
      45200 Develop Backup, Restart, Recovery Procedures
      45300 Define Security Procedures
46000 Training and Cut-Over
      46100 Conduct Training of End Users
      46200 Establish and Test Production Environment
      46300 Go Live with System
      46400 Verify that Business Processes Are Changed
      46500 Eliminate/Terminate Old Business Process and System
      46600 Measure Benefits of New System and Process

Figure 11.10 Tasks for Full Scale Implementation of Software Package

companies have found that it is easier to establish a starting date and enter all transactions into the new system from that date forward. In this case, you might start several months prior to the planned cutover date.

Another major issue exists with history data. Given that the history data is useful to the business, it must be accessible in some form. Business units often want several years of history data. It is infeasible to convert and clean up all of this data, which might require years of research. A frequently used approach is to archive the information to microfilm or to image for access manually later.

Interfaces to Current Systems

Several types of interfaces are required. One type is permanent—that is, the software package will interface with existing software for the foreseeable future. An example might be a case in which the software package will feed an existing general ledger system. There is no plan to replace the general ledger.

The second type of interface is a temporary interface. Here the software package is composed of many large modules or parts. It is impossible to install all of the parts at once. Therefore, the first module of the software package must interface with the existing software, which will eventually be replaced by other package modules.

Interfaces sound straightforward enough, but are they to be on-line or batch? What happens with the conversion of data from the package into the old software and vice versa? How will data be saved from one module of the software package to be input to another module that will be implemented later?

Fit of the System into the Business Process

This has been mentioned repeatedly. The new business process must clearly be defined as being separate from the current process. The project leaders make an effort to begin measurements of the new process and system to validate benefits. Unfortunately, this often means creating shadow systems or making process changes to address the parts of the process that the software cannot handle.

EXAMPLES

SOUTH COUNTY—A NEAR DISASTER

South County is a local government agency that had a number of legacy systems. These systems operated on obsolete hardware. A logical three-part strategy was formulated to achieve modern systems. The first part was to implement new hardware, PCs, and a wide area network. This step was successfully managed. The second part was to select the initial software package that would run on the new hardware, convert over the other legacy systems, and interface them to the package. This step in the strategy fell apart. It cost more than 300% more and over a year longer than anticipated to implement the software package. Conversion and interfaces were pushed back further.

What went wrong? The decision was made to select and implement a general ledger system. General ledger software has unique characteristics that make it very difficult to implement compared to other software. The software requires extensive interfaces to all systems that feed the general ledger. If the old chart of accounts is not adequate, then intensive, detailed work is required to define a new chart of accounts. This is in addition to data conversion. Data conversion often does not work because the level of detail in the old system is insufficient for the new system. The existing legacy systems were based around the old chart of accounts. To have a meaningful general ledger, the old legacy systems had to be modified in addition to interfaces.

What might have been done? Systems that fed the general ledger would have had more users and benefit. The existing general ledger was adequate, but did require improvement. Had South County picked payroll, accounts payable, or accounts receivable, there would have been more users and greater benefit. The general ledger only had 11 on-line users. The other systems had at least three to five times as many.

VIXEN MANUFACTURING

Vixen Manufacturing decided to acquire a large-scale integrated manufacturing system. This system was table driven and required that Vixen operations managers and staff change their processes and procedures extensively to use the software. Vixen management recognized that the operations staff were critical to the success of the project. The managers provided for resources that could help the business units in both current work and conversion. They also indicated to the staff that they were very flexible in terms of rules and policies. They encouraged the staff to first understand the software and then to be creative in their use of the software.

Vixen established a four-person steering committee composed of the business unit, consultant, information systems, and management. The project was divided into subprojects: initial installation and training, business process definition, initial implementation and evaluation, and full-scale implementation. The use of subprojects provided focus to each area of the project. The projects were managed in an open environment. Group sessions were held to simulate alternative new business processes, procedures, and policies.

Even with all of this, there were problems. First, Vixen moved from an initial small pilot implementation to full-scale implementation in one step. This was a problem because not enough people were trained from the pilot to support full-scale implementation. The schedule slipped, and Vixen had to fall back and establish several intermediate stages of implementation.

E-BUSINESS LESSONS LEARNED

Software packages that are included for e-business contain network utilities, firewall software, shopping cart software, credit card processing and authorization, and similar software. Attention is on integrating the software and interfacing it to the legacy systems. New e-business software continues to emerge so that evaluation of the software is more complex. During the evaluation, you want to emphasize flexibility to support marketing and interfaces.

GUIDELINES

- At times it will be possible to keep the old methods in the business process or to adopt new ones that better fit the package. Encourage people to move to the new process. Otherwise, you will see the benefits drip away.
- Identify junior business unit staff members who can be key players in building the new business process and who can act as champions within the business unit.
- Determine at the start of the implementation how knowledge of the software package will be spread out among different business unit staff members. Assume that there will be attrition so that you must allow for backup.
- If you cannot fit a transaction into the software package, be ready to consider a work-around of the software package. If you admit that this is possible early in the implementation, you will likely face more pressure to allow for more exceptions—watering down the benefits of the system.
- During implementation establish an ongoing user group that can be responsible for maintaining the training materials and operations manual and for training new staff in the business unit. This is also intended to head off deterioration of the new process.
- A parallel effort is required on several fronts to reduce the elapsed time of implementation. Each module of the software package must have its own project plan. In this way, there can be at least a limited parallel effort for multiple modules of the software package.

WHAT TO DO NEXT

1. Scan the literature and the Internet to find experiences and lessons learned on implementing software packages. Arrange the lessons learned in terms of management, technology, the software vendor, the consultant, and the business unit. This will assist you in defining roles as well as providing guidance to the people on what can happen.
2. Evaluate your last software package acquisition. Answer the following questions:
   - To what extent was there a complete project plan for implementation?
   - Were the roles of the different players defined at the start of the project or did they evolve?
   - What major issues were encountered in the project?
   - What lessons were learned from the experience?

SUMMARY

Software package acquisition is a growing trend in industry and government. What was once a fairly straightforward project of implementation has grown into a complex one in which you must partner with the software vendor, business unit, and consultant for implementation. This is because business process improvement is linked to the software package. The technical part of the project has also become more complex due to the size of the software package, its interrelationship with the business, customized interfaces with legacy systems, and data conversion tasks.

Chapter 12

Technology Projects

INTRODUCTION

So far in this part of the book, development, maintenance and enhancement, and software packages have been discussed, covering software and other systems projects. However, there are other projects involving networking, implementation of new technology, and so on that do not fit into these categories. These "technology projects" are different from software projects in a number of ways:

- The technology often touches a wider group of users and business unit staff.
- There is potentially more exposure and risk because the organization at large becomes dependent on the technology.
- The projects have substantial costs and in some cases these outweigh software project costs.
- Internal knowledge of the technology is limited because it is new to the organization.
- A wider range of skills is often necessary for successful implementation.
- Benefits are more difficult to pin down because the business processes might not be changed by the technology alone.
- There is substantial, ongoing support in upgrades, operations, and training.
- Technologies may have to be integrated, generating even more work and costs.

Based on these factors it is no wonder that organizations defer technology projects and concentrate on software projects. Upgrades and replacements to current technology are often delayed because management sees the cost, but not the benefits. Given the wide scope of the project along with the cost, the IT organization may view such projects as daunting.

These key questions must be answered:

- What technology projects should be undertaken?
- How do you evaluate and select new technology for implementation given the wide variety of choices and the obsolescence of new technology?
- How should a technology project be managed to minimize total costs over time?
- How can you minimize risk?

APPROACH

YOUR INFORMATION SYSTEMS AND TECHNOLOGY ARCHITECTURE

A systems and technology architecture is the structure of the hardware, software, and network components of your organization. The architecture is more than a list of products and components; it is also the glue and interfaces that tie the parts together. Figure 12.1 gives an architecture for Beaumont Insurance prior to modernization. Note that the architecture even lacks wide area networking. The figure also includes comments on the problems and limitations of the architecture. Figure 12.2 gives the modern architecture for the same firm. This is one of the more extreme examples that can be cited, but it is useful to show you what happens when obsolescence occurs. Maintenance costs of the hardware alone consumed more than $15,000 per month. The mean time to repair problems was in some cases days long due to the problems of the vendor finding staff who still knew the old technology and the difficulty in getting parts.

In comparing two architectures you can begin to see the benefits of a modern, up-to-date architecture, such as the following:

Information Systems Benefits

- It is possible to now acquire a range of modern software packages.
- It is easier to hire staff who are familiar with the new technology than it is to find people familiar with the obsolete technology.
- The dependency on the few critical staff who knew the technology is reduced.
- Maintenance costs may be substantially reduced.
- New software development is now possible.
- Network management and performance are improved with modern tools.
- Operations staffs may be able to be reduced due to central network support.

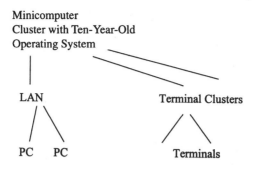

Comments:

• More than 1,200 terminals were connected to the minicomputer systems.
• The LAN and PCs were used only for members of the accounting and information systems groups, not for other users.
• Business units, in desperation, had acquired stand-alone PCs for some functions
• It was impossible to develop new software with the minicomputer systems. No software packages were available.

Figure 12.1   Beaumont Insurance Architecture—Before Modernization

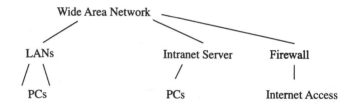

Comments:

- The terminals were replaced by PCs connected to eight local area networks.
- The intranet server supported remote and mobile users, such as claims adjusters and investigators.
- The firewall provides security for selected users to access the Internet.
- Note that processing requirements necessitated the installation of a mainframe computer.

Figure 12.2   Beaumont Insurance Architecture—After Modernization

Business Unit and Management Benefits

- The business can now offer modern and innovative insurance products.
- There is less user training due to working with Windows-based software compared to the old terminal-based systems.
- Employees will have access to electronic mail, groupware, and other types of software that were not possible in the old architecture.
- Reengineering and process improvement can be more easily supported with modern technology.

Are the benefits always achieved? No. One reason is that the new technology raises expectations among business units for more services. Thus, any savings are offset by greater demands for information and processing support. Another reason is that the new technology touches many more people—leading to greater expense.

RISKS IN TECHNOLOGY PROJECTS

You can just jump in and start a technology project. However, that is not a good idea due to the risks and uncertainty. Here are some things that can change or go wrong in a project:

- The number of users of the technology can be underestimated, meaning that a larger investment in hardware, software, and the network will be necessary. This occurs often with intranet and client-server systems.
- The technology does not interface or integrate with existing systems and technology well. This leads to either reduced capabilities or higher cost and more time. This has occurred with some network technologies.
- There was no strategy, so the technology acquired was not complete. This then requires more purchases of additional software and hardware to support implementation. Many technology projects are generated by business unit or management request. Information technology reacts to this without a strategy.
- The situation was not measured before the new technology project. Hence, at the end of the project, no one has a clear idea of the benefits. There is no before and after comparison. This is very common, in that measurement was not planned.
- If not planned, the new technology may not replace the old in all cases. This means that information technology must now bear the burden of supporting both the old and new systems in parallel. This sometimes occurs when a legacy system cannot be totally replaced.
- Staffing support requirements are underestimated. With a fixed head count in information technology, this means that management must play a zero-sum game in allocating resources. This was the case when PCs were acquired in the 1980s. End-user computing support requirements mushroomed.

COSTS AND BENEFITS OF TECHNOLOGY PROJECTS

Technology projects have associated costs and benefits that can be identified now for later reference. Figure 12.3 lists a number of the cost elements of a typical technology project; Figure 12.4 lists benefits. You can extract from these lists the costs and benefits that apply to your specific technology and organizational situation. These are only summary and partial lists.

In terms of costs, network-related components today consume more cost and effort. This is due to the complexity of integrating a number of components as well as linking them to many business units in different locations.

- Hardware and Related Software
  Hardware
  Peripherals: printers, disk drives,
    plotters, tape and backup, and so on
  Network servers
  Print servers
  Database servers
  PCs
  Workstations
  Operating systems
  Utilities
  Compilers and development tools
  Database management systems
  Fourth-generation languages
- Network and Related Software
  Hardware
  Firewall hardware and software
  Routers, hubs, gateways
  Patch panels
  Test equipment
  Monitoring equipment
  Software
  Network management
  Security
  Connectivity
  Monitoring and troubleshooting
  Testing
  Cabling
  Cable interface units
  Telephone and leased lines
  Internet service provider
- Software
  PC software

Electronic mail
Scheduling/calendaring
Internet access
Intranet software
Groupware
- Personnel
  Installation support
    Cabling
    Network
    Hardware and data center
  Systems programming
  Data management
  Network management
  Network support
  Consultant and vendor support
  Training
- Maintenance and Ongoing Support
  Hardware maintenance
  Software upgrades
  Hardware upgrades
  Network upgrades
  Software maintenance and support
- Other
  Duplicate and overlapping support
    during transition
  Facilities work in data center
  Facilities work in business units
  Uninterrupted power supply and backup
  Physical security and access control
  Furniture and office equipment
  Telephone system

Figure 12.3   Examples of Costs Associated with Technology Projects

- Information Systems
  Reduced maintenance costs
  Ability to hire new staff
  Reduced reliance on critical staff with old architecture
  Flexibility in placing software and data
  Increased ability to manage and support the network and users
  Improved security
  Improved management and troubleshooting
  Ability to implement client-server and intranet systems
  Ability to use new development methods and tools
  Ability to acquire software packages
  Potentially less effort in interface support
  Improved productivity with new methods and tools
- Business Units
  Access to new functions (electronic mail, groupware, Internet, and so on)
  Access to new software packages
  Benefits from new development
  Increased sharing of information
  Improved productivity of staff
  Increased computer utilization
  Support for reengineering and process improvement

Figure 12.4   Examples of Benefits Associated with Technology Projects

Note that many of the benefits occur only after application software has been developed or purchased and implemented for the business units. As stated earlier, the benefits of systems generally rest with the business units. However, some benefits associated with increased efficiency and effectiveness are within the information technology unit.

STEPS IN A TECHNOLOGY PROJECT

Given the risks and benefits as well as limited resources, an organized approach to technology projects is needed. Here is a series of steps that can be followed:

- *Step 1:* Determine technology opportunities. This proactive step identifies as many opportunities as possible for new technology.
- *Step 2:* Define the long-term new systems and technology architecture. This is the systems and technology environment that you wish to achieve in a three-to-five-year period.
- *Step 3:* Develop a strategy for technology projects. This involves the sequencing of projects to support the implementation of the new architecture.

- *Step 4:* Evaluate and select the vendors and products for the new technology.
- *Step 5:* Develop the project plan for implementing the technology. This includes measuring the current technology and architecture as well as estimating the benefits of the new technology.
- *Step 6:* Carry out the technology project. The scope includes not only technical implementation, but also use by business units.
- *Step 7:* Measure the results of the project.

In general, you perform steps 1 through 3 one time for a series of technologies. Then you update the architecture and strategy periodically. Thus, steps 1 through 3 spawn a group of technology projects carried out over an extended period. You will initially define the costs and benefits of the new technology in step 1 to solicit management support. These will be refined in successive steps. In step 5 you will determine the costs and benefits in more detail as part of the project plan.

STEP 1: DETERMINE TECHNOLOGY OPPORTUNITIES

The first action in this step is to define your current architecture. This can be done by first listing the various components (hardware, software, and network). Then you can develop network diagrams and text to describe how the components interface and integrate. This is similar to the figures discussed earlier except that you must give more detail on each component.

Now have the architecture lists and diagrams reviewed for accuracy and completeness. As you are doing this, also have people indicate where they think the problems are in the architecture and with the architecture overall. Ask the following questions:

- What components are missing?
- Where are there poor quality interfaces?
- Where is integration needed, but does not exist?
- What does the architecture prevent the business units from doing?
- In what areas is the support effort the highest?
- What is the distribution of costs across the architecture?
- What prevents the architecture from being expanded?
- What is the distribution of incidents of failure in the architecture?

You have a list of issues, problems, opportunities, and gaps in the current architecture. Take a diagram of the architecture and show by arrows and labels where the problems occur. An example appears in Figure 12.5 for Beaumont Insurance. Each of ten areas of concern is described briefly. In real life there were many more problems.

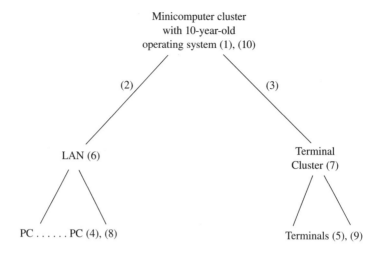

Comments:

1. The hardware and system software are old. Maintenance is high.
   New systems are not possible. Support is time consuming.
2. The LAN connectivity to the minicomputers is terminal emulation and is
   very limited.
3. Electronic mail, intranet, groupware, and client-server systems cannot be supported.
4. The PCs operate in a stand-alone mode with little network use.
5. Users with terminals lack PC capabilities and ease of use.
6. The network within each LAN is very limited in capacity.
7. The cabling for the terminal network is obsolete.
8. The PCs have older software.
9. Terminals do not meet user needs requiring some users to work with
   PCs as well as terminals.
10. Overall, the architecture limits the business.

Figure 12.5   Beaumont Insurance Current Architecture—Issues

STEP 2: DEFINE THE LONG-TERM ARCHITECTURE

You can now proceed with identifying potential components to the architecture defined in the first step. The first action is to determine your overall approach to the new architecture. Today, this typically means that you will identify the following:

- Operating systems for the architecture levels (host, network, and
  PC/workstation)
- General network approach (protocols, connectivity, topology, hardware)

- Scope of the new architecture (which users, business units, and so on will be connected)
- Database management systems/fourth-generation languages to be used
- Development languages, compilers, and so on to be employed

You have now defined the framework for the architecture. The second action is to flesh out the architecture in terms of hardware, software, and network components. Here you will specify each type of component that you require. You will not, however, indicate the quantities required. The results of this action can be reviewed by internal technical staff as well as selected vendors with whom you have relationships and whose products will be part of the long-term architecture. The third action is to determine more precisely the quantities required as well as to estimate the support requirements in terms of staffing.

As the fourth action, you will now develop a list of benefits and costs as well as projects required to move from the current architecture to the new architecture. The costs and benefits can only be rough estimates. With the completion of this step you have defined the new architecture. You are not ready to present it to management, however, because you lack a strategy for implementation. All you will get from nontechnical management now is puzzled support because you have no plan of action.

STEP 3: DEVELOP A TECHNOLOGY PROJECT STRATEGY

A partial list of projects for Beaumont Insurance follows. With the list of projects defined, you must now determine the priority and sequencing of the projects. This is the technology project strategy. It is your implementation strategy.

- Selection and implementation of new host hardware
- Implementation of the wide area network
- Installation of database management systems
- Upgrading and expansion of PCs for business units
- Implementation of electronic mail
- Internet access project
- Implementation of improved local area networks
- Conversion of the existing software applications to the new host hardware

Let's use the Beaumont example to show how to develop the strategy. You first want to determine which projects depend on others. Here electronic mail and Internet access depend on the wide and local area networks. Database management software depends on the new host system as does conversion of the existing software applications.

The second action is to see what can be done in parallel to reduce the elapsed time of installation. Note that you want to reduce elapsed time because the longer the transition, the more expensive and risky will be the projects as you are maintaining the old architecture in parallel with the new. Here you can see that there are three parallel projects: the new host hardware, the wide area network, and the local area networks. You can upgrade the PCs at any time. However, because this involves massive training and support, it can wait. This will be our first phase of modernization. At the end of the first phase, there are few benefits and many costs because you are setting up the structure for the software.

Now assume that the first phase is complete. What is next? Obviously, the priority is to convert the existing application software so that you can eliminate the old hardware. However, access to the on-line applications means that the users must have PCs. The second phase then includes software conversion and PCs. At the conclusion of the second phase, you will have operational systems on the new host and network. Users will be working with PCs instead of terminals. There are some benefits here, but there are even more costs.

The third phase is to establish additional user functions as well as the development and production environment for the future. The projects are electronic mail, Internet access, and the database management software. At the end of this phase you are able to evaluate and select application packages as well as do development.

To summarize so far, the three phases are as follows:

- *Phase 1:* Implement the basis structure of hardware, system software, and the network.
- *Phase 2:* Establish the operational systems on the new architecture.
- *Phase 3:* Implement the development and user functions that take advantage of the new environment.

You have incurred major costs and have received limited benefits. This is a general pattern with technology projects. Even with smaller projects that involve limited new technology, this is often the case.

Now suppose that you have an architecture in place. You can bypass the first two steps to some extent. You will still have to specify how the new technology fits within the architecture. You will also have to develop an implementation strategy for the technology.

At this point you are ready to prepare a presentation and deliver it to management. Many of these presentations end in disaster. Management asks about the benefits to offset the costs. It is likely that you will be unable to identify sufficient benefits to justify the costs. As a result, management turns down the technology project. How do you prevent this outcome? From experience it is recommended that you emphasize the results if the technology project is not carried out. Ask and answer the following questions.

- What will happen to the architecture if deterioration continues over several years?
- What is the impact on the business units if modernization does not occur?
- What will be the comparative position of the organization in industry?
- What are competitors doing with technology? What can the competitors do that you cannot do with your current technology?
- What is the impact on the ability of information technology to respond to business unit requests?
- How will the business units benefit from the new technology in the near term?
- How will the business units benefit from the new technology in the long term?

This approach emphasizes the downside of continuing business as usual. It is the key to getting management support. At the end of the presentation you will indicate that the next steps are to identify precise quantities of each component and develop a project plan. You will then return to management with the results of these two steps. Although you can do these steps first and then present the whole thing, it is not a good idea. Use this presentation as a means of showing management a vision of the future.

STEP 4: EVALUATE AND SELECT PRODUCTS

Prior to this step you have gathered some information on products. For database management systems you may have narrowed the field to two or three products, for example. Create a project plan to perform the evaluation and selection. Because evaluation and selection were already discussed for software packages, the attention here will be on what has to be included in evaluation and selection.

Involve the business unit managers and staff in this step and those that follow. The business unit will have to support the selection as well as the plan and benefits. If the business unit is not involved here, but only later in implementation, then there is a risk that the project will be labeled as only an information technology project. This can doom management support when you present the plan.

You must ensure that the lists of components and support tasks are complete. If you miss something important and have to return for approval, you risk the entire project as well as your own credibility. Make the effort to contact existing users of the technology and find out what components they acquired, their experience in implementation, and some idea of the costs incurred and benefits received.

You can contact the vendor at this stage to review the list of components and to get ideas related to the implementation plan for the technology. It is important that you get the vendor on board because the vendor will later be asked to provide support. Through the vendor's participation you protect yourself in terms of being complete.

STEP 5: DEVELOP THE PROJECT PLAN

Although the project plan for a technology project depends on the specific technology, there are some general stages of work. Divide the project plan into these stages:

- *Stage 1:* Make preparations for installing the technology. This might include initial training in the technology, facilities work, staff assignments, procurement, and related start-up tasks. At the end of this stage, you are prepared to install the technology. Here you will also measure the current business process and technology with the business unit.
- *Stage 2:* Install the technology. This is followed by testing to ensure that it is functional. Of course, it is not connected or integrated with anything as yet.
- *Stage 3:* Employ the technology in a prototype mode. This includes the initial loading of data, establishing some of the interfaces with the technology, and starting on any customization. During this stage both information technology and business unit staff members will become familiar with the technology. Benefits can be estimated again.
- *Stage 4:* Complete the implementation of the technology. The technology is used in production. The results of using the technology are measured and compared with the estimated benefits.

This stagewise approach has several benefits. First, each stage has defined, measurable milestones. Second, mistakes or errors can be addressed prior to full-scale implementation and use. Third, employees are given time to work with the technology. Fourth, you get an opportunity to show managers what is going on and how you are protecting the business unit from any unforeseen problems.

It is important to have both the business unit and vendor involved in project planning. This joint effort will build participation, commitment, and support for the project and the technology. The business unit will have to make the technology work within its business processes.

It is now time to review the selection and plan with management. You will want to focus on the business and management aspects of the plan and the technology. Thus, you focus on how the technology can be used in the business processes and what benefits it will provide. You will not focus on how it works or the details of the technology. Have the business unit manager give the part of the presentation dealing with use of the technology and its benefits.

STEP 6: IMPLEMENT THE TECHNOLOGY

The implementation will proceed following the stages of the plan. Consider having a separate project team for each stage because the major tasks in each stage are different. Business unit involvement in the first stage is limited to learning more about the technology. There is no or little involvement in the second stage of installation. Involvement grows in the third stage. Several people from the business unit will be involved. They will have to prepare operational procedures for the technology, define how it is to be managed, support the conversion, and perform testing. The business unit role expands sharply during the implementation stage.

Technical staff such as systems programmers and network staff are involved in supporting the installation in the second stage. Programmers and analysts take over for the third and fourth stages. Vendors support the first three stages.

Beyond the project organization, it is very important that you select the right first application of the technology. If you pick one that has time pressure, then you risk failure because the project team is just learning the technology. If you select an application that involves a business process that is in trouble, there is also unnecessary risk. The critical lesson learned is that you select the application that (1) allows team members ample time to learn the technology, (2) presents little time pressure, (3) offers clear benefits to the application, and (4) is arranged so that the business environment of the application is representative of the potential applications.

Picking the first application is part of your implementation approach. It is hoped that you will have the same project team members available for the second and third applications of the technology. Choosing the first application was based on minimizing risk and maximizing the chances of success. Selecting the second application must be based on benefits. Therefore, the second application typically involves more users and a greater number of interfaces.

After each application of the technology, measure the project as discussed next. You will also want to gather lessons learned so that your plan, issues, and how you approach situations will be more effective as you go. In one large implementation of fault-tolerant computing, there were eight different applications of the technology. The second application produced more benefits due to a wider scope. Lessons learned and new experiences were gathered during the first four implementations of the technology. It was only during the fifth implementation that little new was learned.

STEP 7: MEASURE THE PROJECT RESULTS

The scope of the first stage includes measuring the current business process and system. The last stage encompasses the measurement of the business process after the technology has been implemented. In this step you will also gather lessons learned from the information technology and business unit staff. Some areas of lessons learned will bring forth the following questions:

- What surprises were there at each stage?
- How did the business unit ensure that the benefits were achieved?
- What technical tips were gained as a result of the experience?
- What experience and knowledge were gained from the implementation and use of the technology?
- What new information was gained on the condition of the old business process?

When implementing a new technology, the current business process is often streamlined and reengineered to take advantage of the technology. This work, begun in stage 3, provides you with insight as to the detailed problems with the current process. The information technology staff may also have ideas about how to further improve the architecture to take advantage of the technology.

In documenting the measurement, begin with describing in detail how the current and new business processes work. Then you can show the differences before and after. Follow up by identifying the areas of benefits and costs associated with the project. You will change the order when you present the results to management. During the presentation, the business unit manager will show the current process and highlight the benefits of the technology. If necessary, you and the business unit manager can illustrate the difference between the old and the new by showing a sample business transaction that takes advantage of the technology.

During this step, also try to identify the most efficient and effective means of achieving ongoing support for the technology. When people installed PCs for the first time on a wide scale, they were surprised by the quantity and persistence of the end-user computing support required. New users needed not only initial training, but hand holding as well.

THINGS THAT GO WRONG
AND WHAT TO DO ABOUT THEM

Many things can go wrong in technology projects. Here is a sample of some of the potential pitfalls, gathered from experience, observation, and war stories.

- *Midimplementation paralysis.* In this case the technology was successfully installed the first time. However, due to the problems encountered it was decided to eliminate or delay any further implementations. This means that the economies of scale of support for the technology on an ongoing basis are spread across a smaller number of users. This occurred with some applications of database management systems.

- *Niche technology.* The systems group or business unit identifies a technology that has very limited application. It is successful, but it consumes too many resources to do the implementation. This occurred with early implementations of image technology.
- *Failure without backup.* One company tried to use digital cameras to capture photos and then transfer them over the Internet. The quality of the pictures was not acceptable and the camera did not hold enough pictures. The company dropped the project, wasting both money and a good idea. The company might have gone back to using standard 35-mm cameras with scanning but did not consider it.
- *Business unit is difficult to work with.* Information technology finds that the business unit staff lack interest in the technology and therefore fade from the project. Instead of facing this problem head on, IT pushes ahead to implement the technology, based on the belief that when the users see the technology installed, they will like it. This is unlikely, however, given the users were not involved.
- *Too many technologies in one project.* Many early client-server projects failed because their scope included middleware, client, server, and host interface development. The project was too ambitious.
- *The technology has just appeared.* The early personal digital assistants (PDAs) were often failures due to their limited recognition ability and processing power. Applications based on these failed. You want to select technologies that are sufficiently mature for industrial strength use. How do you know when this is the case? When other companies have applied them. Don't be a pioneer. You can end up with arrows in your back.
- *First application requires extensive integration with legacy systems and data.* This project will turn into one that is 90% legacy system programming and 10% new technology. Many early efforts at work-flow software and client-server systems have suffered this fate.
- *Implementation gets out of control.* The initial application of the technology was extremely successful. Rather than measure the results and regroup the resources, the team is forced into a large project. The team encounters too many problems. The project is stopped and termed a failure. This occurs when a limited pilot project is followed by a desperate desire to implement fully.
- *Selected technology is undergoing rapid change.* In the early 1990s, network technology and operating systems were undergoing change. This accelerated in the mid-1990s. As a result, some companies that plunged in to get a network operational ended up having to redo the project based on the latest technology. Select technology that will be stable over some payback period.

EXAMPLES

ELECTRONIC COMMERCE

Electronic commerce is a rapidly growing area of business for retailers as well as distributors. It sounds so simple. Get some electronic commerce software with some hardware and you are in business. This is a myth. Here are some of the questions that you must answer when embarking on an electronic commerce project:

- What are the benefits to electronic commerce? How much will sales increase or costs decline?
- Who will be your initial partners in using electronic commerce?
- What will be your strategy in terms of the expanding use of electronic commerce?
- Electronic commerce typically requires that you develop and implement new on-line interfaces between systems. How will this be handled in the short term given the effort to interface with legacy systems?
- How will internal business processes be affected by electronic commerce? Will you have to establish parallel processes for electronic and nonelectronic means?

GROUPWARE

Many people use groupware. The more people that you can link together, the more successful groupware will be. Groupware also has many possible applications. It is fairly easy to use and applications can be developed quickly. So why do groupware implementations run into trouble? Here are some typical issues:

- Groupware products often lack a database management system structure. This tends to limit the applications that are possible in terms of programming business rules.
- Interfaces to host systems are often not possible with major programming effort.
- Many people resist groupware. They would rather continue with what they are using and "what works for them." This may be telephone, fax, memo, or electronic mail. A major challenge in groupware is to establish a sufficient mass of applications to encourage use based on self-interest.
- How do you measure success? If people use groupware, what did it replace? Because groupware often consists of a large group or set of applications of occasional use, measurement is a challenge and may be addressed early in the project.

A major communications company implemented groupware across Latin America for its operations. It only took three months to install the system, train users, and develop applications. The company sat back and waited to see what would happen. There was very little use. Management eventually mandated groupware use and banned faxing and paper. This ensured success. Thus, it is important to determine the policies that will be necessary to accompany the technology.

INTRANET/INTERNET

A number of the issues that companies are struggling with impact the projects that use Internet technology:

- To what extent will employees have access to the Internet? What are the benefits from such access?
- How will the Internet use be measured and cost ustified?
- If there is to be an internal intranet, what types of applications will be supported?
- When will an application be implemented using intranet versus a client-server system?
- What is the order of business applications? Should there be many applications for a smaller group of users, or should there be fewer applications for many users?
- How will users be supported?
- What is the impact of an intranet on the PC software that users have and how will it affect their training?
- Who will set up and maintain Web pages? Who will maintain Web content? If pages are not updated, what will be the criteria for killing off pages? How will usage be measured? If Web pages are employed, then will information still be available in traditional forms?

E-BUSINESS LESSONS LEARNED

Technology implementation is a key component to e-business implementation and support. You have to upgrade the network, provide additional computing capacity and servers, provide for backup and recovery, support testing and quality assurance, and move to a 24-hour-a-day/7-days-a-week support environment. Keep in mind that the technology must be scalable. The spread between high and low levels can be hundreds of percentage points over the low level.

GUIDELINES

- As technology changes, you might be tempted to undertake many technology projects to keep up to date. This is the opposite extreme to doing nothing. Too much technological change is disruptive and counterproductive. Moreover, it drains resources from other projects and halts projects while the technology is being put into place.
- Consider the long-term use and potential application of any new technology. This will aid in determining your strategy for the technology.
- Many organizations underestimate the cost and effort of getting rid of an existing technology when implementing the new one. It is not just conversion, but also training, interfaces, and an extensive testing and shakedown effort that are required.
- If your legacy system depends on obsolete hardware or software, then you must consider replacing the legacy system as you replace the hardware and software.
- Major costs of obsolescence lie in ongoing maintenance as well as in the declining staff morale that can arise when people are surrounded by dinosaur products.

WHAT TO DO NEXT

1. Answer the following questions about your current systems and technology architecture.
   - Do you have a defined architecture that employees are aware of?
   - Is there a formal approach for evaluating and selecting new technology?
   - What cost-benefit analysis is performed for a new technology?
   - Are technology projects treated and managed differently than software projects?
2. Based on your knowledge of your organization and its technology, identify areas where the architecture must be modernized. What are the tangible benefits of modernization? In implementing the technology, what interfaces are necessary?
3. Find some technology that has been in limited use but that had been projected to be used on a wider scale. Why wasn't its use expanded? What are the ongoing support costs and issues associated with the technology being limited to a few users? How does this technology lock its particular users into limited future change? Is it better to continue using the technology, or should it be replaced?

SUMMARY

Your systems and technology architecture is a dynamic structure that requires you to spend time periodically to consider improvements and upgrades. If your technology becomes obsolescent, then you will have likely been blocked from acquiring new software packages and carrying out much new software development. In other words, a poor architecture dampens your ability to improve your business processes. Moreover, the more you fall behind, the more expensive and time consuming it will be to catch up. Technology projects have long-term structural goals for the architecture as well as near-term goals to support the organization.

Part IV

How to Successfully Address Project Issues

Chapter 13

Business Issues

INTRODUCTION

This chapter, and the others in this part of the book, begins by introducing each issue. This introduction is followed by a discussion of factors that give rise to the issue, the impact of the issue if not treated, how to prevent the issue from occurring, and what to do if the issue arises. The chapters are arranged slightly differently from other chapters in that the examples and guidelines are included within the discussion of individual issues. How can you use this material? The first answer is obvious—employ it when the issue arises. A second answer is to create a checklist.

As systems organizations direct their resources toward client-server, intranet, extranet, data warehousing, and other applications that closely touch the business, the number and extent of business issues will continue to rise. Unlike personnel or technical issues, business issues force the project leader to move outside of the project team and deal with managers and employees in various departments. To head off problems here, it is likely that the project leader will spend even more time with the business issues and away from technical areas of the project.

ISSUES

ISSUE: THE BUSINESS UNIT CHANGES REQUIREMENTS FREQUENTLY

This is one of the most common complaints voiced by systems managers and staff. It has been so for the past 30 years and is not likely to disappear soon.

How This Occurs

Are the requirements really changing? If the requirements are based on transactions in a business process, in management, or with external forces, then requirements have changed. If the source of the change is what a middle manager might like, then it may relate to style or politics rather than substance. A manager can even seek change for political gain or to delay the project—these are sinister, but possible sources of the change. There can also be a lack of understanding of systems and technology.

Potential Impact

A change to the requirements is often viewed with alarm by the project team. Team members see their work being undone and, in the worse case, fear having to start over. This can be frustrating and demoralizing. This situation is made more severe if people have experienced requirements changes repeatedly on multiple projects. They may just become resigned to having it happen again.

How to Prevent the Problem

To prevent this from happening, the systems group must become more proactive at the start of the project. The scope of the project must be negotiated. This is important because it is one of the leading areas where requirement changes occur. The second step is to base requirements on the definition of a new business process. Requirements based on detailed business transactions and rules are less likely to change. Any change in requirements can affect the business process. The business unit staff must be heavily involved in defining, understanding, agreeing, and carrying out requirements.

How to Address the Problem

If a requirements change occurs, answer the following questions:

- Why did the change appear now?
- What is the impact if the change is not addressed?
- What is the impact of the change on the underlying process?
- Can the requirement be met through procedure or policy change rather than a systems change?

Consider reopening all requirements and specifications with the business unit. Additional requirements can be identified now to minimize impact on development.

ISSUE: THE BUSINESS UNIT DOES NOT PROVIDE GOOD PEOPLE FOR THE PROJECT

Business units today do not have a surplus of people. Many departments have downsized. As a consequence, only a few people have in-depth knowledge of the business rules and transactions. Others may know what to do, but they do not understand why they are handling the transactions in that way.

How This Occurs

When asked to provide someone for a project, a business manager cannot afford to release the most experienced person. The department might really suffer. In addition, the manager may lack confidence in the project and might not want to risk good people on the project. Managers are often likely to provide individuals whose loss might have the least impact. They may have had poor project experiences in the past. Timing may be bad—the change might be conflicting with the year-end closing, other projects, or a peak work time of the year.

Potential Impact

If you assume that the business people provided to the project can speak for the department, you are likely to be sorely disappointed. Getting information from them without validation and checking on completeness can be a problem. The team might carry out work that will later have to be redone.

How to Prevent the Problem

Ask for someone on your team who is junior in the business department and request only their part-time involvement. Assume that the person has a limited knowledge of what goes on in the department. Request that a group be appointed to review the work for validity. If this is not done, then find other people to review the work. You are picking a junior person because that person's absence will least impact the business and because you really do not know what you need yet. This has the side benefit of gaining broader participation in the project from the business unit. You know that you understand the unit's limitations.

How to Address the Problem

If the business unit provides no one, then question the project. If the business unit is uncooperative at the start of the project, then things will likely go downhill from there. Meet with the business manager to indicate your minimal known requirements and your sensitivity to his or her staffing problems. Also, point out that because the project is just starting, you really do not know who and how much of the person's time will be required.

ISSUE: THE BUSINESS UNIT IS UNWILLING TO CHANGE THE BUSINESS PROCESS

How This Occurs

Members of the business unit can become convinced that the way that they are handling their work is the best and only way. After all, it has been proved through years of use. They may also fear change and the impact and disruption likely to follow. They may feel that any change will make matters worse. They were promised much by information systems and may feel that they received little. Middle managers may resist because they fear a loss of jobs and power.

Potential Impact

Resistance can be open or subtle. Delays in making decisions are one sign of problems. Change of mind and indecision are other signs. The project now suffers due to a lack of defined specifications. The business unit employees may keep trying to bring the project back to the current system.

How to Prevent the Problem

What was said in earlier chapters applies here. The scope of involvement and role of the business unit have to be nailed down in the definition of the project concept. This has to be reinforced as the project proceeds. Not only must the benefits be stressed, but the problems must be stressed also if the current process continues.

How to Address the Problem

If the business unit wants a new system and also wishes to maintain the current process, then you must answer some fundamental questions:

- What is the benefit of the system if the process is unchanged?
- How will exceptions not handled by the current system be addressed by the new system operating with the old process?
- If the business unit won't change the process, what sign is there that it is even willing to accept a new system? Won't the business unit want the new system to be just like the old one because it fits well with the current process?

Show how the new system fails to mesh with the current process—creating more problems.

ISSUE: SEVERAL BUSINESS UNITS CANNOT AGREE AMONG THEMSELVES

Major systems projects with substantial benefits often cross multiple departments or divisions, making this issue more common.

How This Occurs

Many departments have never really worked together on a systems project. Each brings its own history, opinions, and preferences into the project. Each department's relative power structure also comes along. This makes for an impossible situation if the project manager is naive. The same can be said of projects that involve multiple offices spread across different parts of the country or the globe.

Potential Impact

The project can become paralyzed because no decisions can be reached. Any decisions that are achieved may be undermined by departments who feel that they lose power if the decision appears to go against them. If hostility continues to rise, it may be impossible to continue the project without major and ongoing involvement of upper management.

How to Prevent the Problem

Try to understand the politics of each department at the start of the project. Meet with business managers and identify the potential problem of generating consensus. Seek agreement on a method to resolve problems in advance. Identify potential conflicts with the project team at the start. Show how the approach and solution fit within each department's self-interest.

How to Address the Problem

If a problem arises that is due to disagreement, then work with each department individually to get opinions and concerns. Don't address this in the group at first—it will likely solidify positions. Develop a solution and interpret it from the perspective of each department's self-interest. Follow this up with additional sales efforts. Take this problem seriously. Such issues can appear trivial to an outsider, but they are not to the people in the trenches.

ISSUE: THE BUSINESS UNIT CANNOT CONSISTENTLY RESOLVE ISSUES

You thought you had resolved an issue, but each time you turn around, the situation is changed and is still unresolved.

How This Occurs

Indecision may be due to lack of understanding of the issue and its consequences. People in the unit may not be prepared to make decisions. Alternatively, there may be a power struggle with different factions within the department. A fourth possibility is that people in the unit do not know their own business process in sufficient detail. The situation can be made more complex if there is lack of communications between levels of the organization.

Potential Impact

The project may become frozen. The systems people may throw up their hands and state that they will await the final decision. Morale in the team may drop. If the decision is forced, the possibility of it resurfacing remains.

How to Prevent the Problem

Issues must be analyzed in terms of the basic business process, which is more likely to yield truth than opinion. Any resolution based on the process is more apt to stick. However, you will then have to provide a political slant on the decision for each department. Don't wait for the department to point out the problems. Be proactive. Each new issue is to be reviewed in terms of what has already been resolved to see if it is a variation.

How to Address the Problem

Let's assume that you get conflicting signals. Consider the issue again along with all outstanding issues. Use this as an opportunity to review the roles and responsibilities of the department as well.

ISSUE: THE BUSINESS UNIT STAFF MEMBERS DO NOT KNOW THE BUSINESS PROCESS

How This Occurs

On the surface, this sounds absurd. However, today, with downsizing and change, it is true more often. The people who possessed the deep business knowledge may have retired or disappeared. Although individuals know pieces of a business process, no one knows the whole. In addition, people do not know why they perform the work the way they do. There may be a difference between the way people think that the process is being done and the how it is actually being performed. Knowledge may be fragmented through the group. There was once a soldier who appeared next to a bench and guarded the bench from 11:00 A.M. until 2 P.M. When asked why he did this, the soldier replied that he didn't know, but that it had been going on for years. When a probe was initiated, it was found that four years before, the bench was painted. An officer who failed to notice got wet paint on his uniform and then ordered someone to guard the bench while it was wet. The order was never rescinded.

Potential Impact

You cannot assume that business knowledge or information is complete or accurate. This can also lead to problems in specifications and design. When there is a lack of information, there is the danger of a process being invented from a technology view without business involvement.

How to Prevent the Problem

It is best to observe the business directly and see the situation first hand. Ask to be trained as if you were a new employee. How many people does this process take? Are the instructions incomplete or contradictory? How often do you have to ask for assistance? Doing this will provide you with contacts as well as information on the state of documentation, the extent of staff knowledge, and the existence of formal procedures.

How to Address the Problem

Requirements may be provided by managers and staff. However, they may be incomplete and fail to address exceptions. As you are provided with these requirements, verify them with the business process directly.

Issue: The Existing Business Process Is in Poor Condition

A business process can deteriorate like a system over time. More work may be handled informally outside of the process. The error rate may be high. Transactions may have to wait for a specific person to handle them due to overspecialization. The supervisors may be handling a substantial percentage of the work themselves. The department may be using homegrown PC-based solutions that are incompatible with the major systems.

How This Occurs

Deterioration in a process often sets in because people take the process for granted and fail to give it any attention. No one thinks to put money in the budget to fix the process. After all, processes fix themselves right? Once the situation worsens, managers may be reluctant to address the problem especially if the process works at some minimal level. They fear upsetting the department and making the situation worse.

Potential Impact

Any system linked to a deteriorated process may suffer from faulty data, dis-use, or abuse. The system may not be employed as intended. A new system has a small chance of success if the process problems are not addressed. You cannot as-sume that the new system will fix the problems of the business process.

How to Prevent the Problem

Investigate the state of the business process as part of the initial work toward a new system. Include the process within the scope of the project. Measure and doc-ument the condition of the process at the start. Separate out what can be handled in a system rather than the process.

How to Address the Problem

If work has begun on a new system without an in-depth review of the process, your project is in trouble. Validate the design against the process. This will draw attention to the condition of the process.

Issue: Middle Level Business Unit Management Resists Change and the System

How This Occurs

Who is threatened by the new system? People who work with the current process and system may be terrified, but they have little to fear. Transactions will still be processed—only with a new system. More directly threatened are man-agers above the immediate supervisors of employees who perform work in the process. If not brought into the project and without a future role being defined, they can become resentful and disruptive.

Potential Impact

Lack of cooperation by mid-level managers can create problems at lower levels. It will take longer to gain support and reach decisions. Project progress will slow down.

How to Prevent the Problem

Planning ahead, you will want to work with upper management in the business unit to gain support and to ease any fears of change. If some managers are to be released later, then the project team can be guided around these people. When asked, the team can indicate that the scope of the project includes the process and system, but not the organization. It is best if upper managers identify their future roles to gain their participation.

How to Address the Problem

If you encounter resistance from mid-level managers, then you can take several courses of action. One is to sympathize and indicate the project scope. Another is to minimize contact so as to prevent confrontation. In the most serious cases, you can contact upper management in the business unit. Push for role definition. Do this rarely as you don't want to be labeled as someone who "cries wolf."

ISSUE: THE BUSINESS UNIT ATTEMPTS TO DOMINATE THE PROJECT

How This Occurs

Business managers who realize the importance of systems in relation to the business process may seek to dominate the project. This is due to a desire to craft the new process and to fix errors. It may also be a defensive move to retain control. They may also fear the project results if they release control.

Potential Impact

Although the initial impact may be positive to give more support to the project, the longer-term effect may be more changes in the project due to meddling. There can be an attempt to retain the current process.

How to Prevent the Problem

The situation may be prevented by defining the business role at the start. This can be followed up by having the business unit involved in resolving issues related to the project. Once implementation has begun, the business unit will, after all, play the major role. This might make the business manager hold off from over-involvement.

How to Address the Problem

If domination occurs, don't discourage it. Try to channel it toward issue resolution. Define roles by assigning areas of tasks to the business unit.

ISSUE: THE BUSINESS UNIT VIEW DOES NOT FIT WITH THAT OF UPPER MANAGEMENT OR THE INDUSTRY

How This Occurs

You have approval for the project. Management sees the project as critical for success in competition. Yet the business unit takes a local view for its own interests. The two views are dramatically different and opposed. This is not unexpected given the mission and charter of the business unit. The business unit may also be out of touch with trends and practices in the industry today.

Potential Impact

Left unaddressed, the clash in purpose and scope can harm the project. Relations between the business unit and management may deteriorate. Upper management may not want to take on the department so that the project is either warped or flounders. Alternatively, there may be no surface conflict, yet conflict does exist.

How to Prevent the Problem

The issues of purpose and scope must be handled in the project concept. This will give management a forum to direct the business toward a wider goal. An effort must be made to relate the wider goal to the operation of the business unit.

How to Address the Problem

Resistance to a larger project can still emanate from lower levels of the business unit. The project team acts as a marketing arm of management to reinforce the wider project view and to show how new work processes and systems support both the department and management views.

ISSUE: BUSINESS UNIT MANAGERS AND STAFF LACK TECHNICAL KNOWLEDGE

No one expects business managers and staff to be programmers or analysts. However, lack of knowledge of basic technology concepts can impact a project because it makes the fit between the system and process more difficult.

How This Occurs

Individuals can go to school and then work in a department for many years using technology. However, they may lack awareness of its potential and may not be up-to-date with the technology. This is not a surprise because the technical staff members lack knowledge of the business department. In such situations many departments just accept the technology with which they are provided. The business unit staff members may not see how their jobs might benefit from training in the technology.

Potential Impact

Though there is acceptance, employees may be reluctant or lack the creativity to develop new ideas for systems improvements. When systems staff ask about new requirements, they may be met with blank stares. Employees may not comprehend how the system will provide benefits.

How to Prevent the Problem

You are not expected to train everyone in technology. What you can do is explain new technology in terms of the information on the business process that you have collected. Staff members are more likely to be interested because it is relevant to them. This can stimulate their creative thinking as they respond. You can treat the technology as a "black box."

How to Address the Problem

When you encounter resistance to technology, it may be due to ignorance and fear. Do not assume hostility. Take time to explain the technology in terms of how staff members can use it and what its benefits might be. Don't explain the details. Show that it performs similar functions only better.

ISSUE: BUSINESS UNIT MANAGEMENT IS REPLACED

How This Occurs

In any project with an extended life span, it is possible that key business unit managers who provided support to the project may be promoted or leave. Change can occur through reorganization, mergers, and so on as well.

Potential Impact

The impact of management change is often viewed negatively. People want to put the brakes on the project until things settle down. This can cause the project to stall. By putting a positive spin on the impact, you create an opportunity for new support and fresh ideas.

How to Prevent the Problem

If your project lasts a year or more, then anticipate management change. Market the project concept to as wide a group of managers as possible. Keep them informed and involved. The negative impact caused by the departure of one manager is minimized. Within the business unit build project support at the grass roots. This will come in handy in dealing with new managers. Make sure that you work quickly to bring new managers on board. Get them involved in issues that are not time sensitive so that they can adapt to the project and adopt it as their own.

How to Address the Problem

If you encounter sudden management change, establish communications with the new manager. Don't wait for the manager to come to you. Point to the progress and benefits as well as the manager's role in the remainder of the project. Involve managers in issues.

ISSUE: THE BUSINESS UNIT HAS NO INTEREST IN THE PROJECT

How This Occurs

Even if something is in your best interest, you may not have the time or energy to devote to it. It is that way for financial groups during the annual year-end closing of the accounting books. It is the same with shipping and distribution firms during peak periods. Alternatively, people in the business unit may be keeping their eyes on their immediate issues and fail to see the benefit of the project even though company management wants to see the project completed for overall benefit.

Potential Impact

Lack of interest can spell doom or delay for a project. If you attempt to reverse but you fail, the project can be cancelled. Upper management support for the project can be countered by tactical arguments, such as that there is too much work that is more important than the project. This leaves upper managers with few options—they will delay the project. To understand the impact, you must understand the cause.

How to Prevent the Problem

During the initial definition of the project, point out the project benefits. Doing so will appeal to the self-interest of the business unit. If the senior management of the business unit does not show interest, move down in the organization to obtain support. If there is little benefit to the business unit directly, then discuss the potential problem with upper management to get ideas of how to address the issue before a lack of interest arises. You seek to establish support for the project at the top and the bottom.

How to Address the Problem

Suppose that although members of a department do not show interest, there still are project benefits. Consider involving the departments that will benefit, and get them to help you bring the reluctant department along.

Management interest in the project may also wane over time. This might be due to problems in the project or in the perception that the benefits will be less than anticipated. Revisit the project concept and determine if you really want to go ahead or change the project.

ISSUE: OTHER WORK OR PROJECTS HAVE HIGHER PRIORITY FOR THE BUSINESS UNIT THAN YOUR PROJECT DOES

How This Occurs

Your project is underway. People assigned by the business unit are reassigned to other work. They disappear. Priorities can change. Problems can arise in the department that require the specialized skills of the team members. Remember that in many cases in business units, the expertise of all of the trained staff is required to address all of the exceptions.

Potential Impact

Without any notice, the project can be stranded in a vacuum. Questions will soon arise that require answers from the business unit. If you hassle the people who returned to the department, they may resent the additional work and may not want to return to the project. Although parts of the project can proceed at a slower pace, there will be a noticeable effect as work is deferred waiting for answers.

How to Prevent the Problem

At the start of the project, never ask for people from the business unit to be assigned to the project on a full-time basis. Try to involve a number of people so that you are not overly dependent on one or two. Keep in contact with the business unit manager so that you know what pressures the unit is working under. If you sense that the team members are required back in the department, work with the business unit directly to plan on how to cover the gap. In a project that will last from six to nine months, plan on work of higher priority arising and attempt to schedule it in the project.

How to Address the Problem

If there is a sudden crisis and departure, expand your network of contacts in the business unit. Develop a contingency plan to keep the project going prior to contacting the business manager about the problem. While indicating the impact, sympathize with the manager.

ISSUE: THE BUSINESS UNIT HAS EXISTING TECHNOLOGY THAT CONFLICTS WITH THAT OF THE PROJECT

How This Occurs

There is probably technology in your home with which you feel very comfortable and that you don't want to replace. It is the same in business. A department may like its procedures, policies, and systems. People in the department don't see a need for change. They may have created PC solutions that they don't want to lose.

Potential Impact

Not only can the project be derailed, but you can encounter long-term, deep-seated resistance. You may fight a constant uphill battle as the business staff members find numerous holes and gaps created by the new system replacing the old.

How to Prevent the Problem

When you begin the project, be sensitive to people's feelings. Indicate that any new system or technology can build on the experience and knowledge associated with the existing system. Also, show them that many procedures and policies can be reused. The new system is then an evolution from the current system. Reinforce this benefit by focusing on the additional features and capabilities of the new technology and system that are lacking in the current system. Show how the functions performed by PC and other systems are addressed in the new system.

How to Address the Problem

If you sense resistance, then expand your contacts in the business department. Take more time to understand how the current system and process work. Ferret out the PC systems. Lower the profile of the project to avoid being a target. Disseminate information about the new system to show how the new system is better than the current one.

WHAT TO DO NEXT

1. With your list of issues in hand, separate out the business-related issues. Look for patterns in the issues. Review your relationship with all levels of the business unit to identify potential actions you can take.
2. Taking a wider perspective, review the history of the project to assess how business-related issues were handled. Add this analysis to the results of item 1.

SUMMARY

Constantly monitor the relationship between the project and the business units involved in project. This is not only to detect potential problems, but also to identify opportunities for increasing the involvement by the business unit as well as its participation in addressing issues. The importance of the business unit contribution to the project can be highlighted by indicating results from the involvement.

Chapter 14

Human Resource Issues

INTRODUCTION

Personnel issues in projects were of concern in ancient history, prior to the formal definition of projects and project management. Attention and concerns have grown due to a number of factors:

- Downsizing has forced companies to do projects with fewer people so that the performance of the staff on the project becomes critical.
- Deadlines for systems and technology projects are tighter along with budgets as management realizes the role of systems in competition and business operations.
- Business processes depend on systems more than ever—raising the level of the importance of projects.
- People are shared among projects, creating more issues and tension in managing human resources.
- It is more difficult to attract and retain good people.

Human resource issues focus on the project team. The project team may include staff from business units as well as internal systems staff.

ISSUES

ISSUE: TURNOVER OF PROJECT TEAM MEMBERS

Let's begin by acknowledging that some staff turnover is expected in a systems project that spans a year or more. In addition, it is sometimes favorable that certain individuals leave the project team. Thus, turnover is not bad in general. When a person leaves the project on schedule and as planned, then it is not counted as turnover. However, if the extent of turnover is great, or if critical people leave, then there is a problem. Note that the approach of having a small core team with many part-time people or players helps mitigate the turnover problem, except within the core team.

How This Occurs

Individuals can get tired of the project or begin to feel that their work is not appreciated. There may be many issues in the project. People may feel that because the project is in jeopardy, their jobs are at risk. Team conflicts and personality clashes can be another cause. Because many systems people are often quiet and not outspoken, it is sometimes difficult to detect problems among them until it is too late. The team member then seeks to return to the line organization and be assigned other work.

An alternative reason for turnover is that a person (including even the project leader) gets an attractive offer either internally from another department or externally from another firm. The individual may discuss it with his or her line manager. It may not even be brought up to the project leader if the leader and the employee lack a close relationship. If the project leader is not in regular and constant contact with the project team members, then many times the situation cannot be turned around. The person leaves.

Potential Impact

Damage occurs if the project manager is caught off guard. Then the leader does not look like a leader. Faced with such a problem, you must scramble around to find someone else and to carry out damage control within the team and with management. The problem is compounded if the team member involves the line manager. The line manager may question the project leader and the leadership itself. It will obviously be more difficult to attract a replacement into the project.

If the person is hired from outside, then another set of problems emerges. The remaining team members may feel that they can do better outside. The departure of one person may lead to a rout. This is often precipitated by headhunters and search firms who attempt to systematically raid an organization of its technical talent. This results in more pressure on the team.

How to Prevent the Problem

The project manager must perform several important activities. One is to stay in close contact with team members. The project leader must establish sufficient rapport so that the team member will alert the manager to problems and to headhunter activities. As a project leader, you must keep such information to yourself. You cannot attribute it to the particular individual; otherwise, you affect the communications. The person may not take you into his or her confidence again.

Ask if the team member has any concerns or problems with the project or the team on a regular basis. You can do this in a positive way by asking if the person has any ideas that might improve the project. The ideas you solicit may lead you to the problem (reverse engineering).

On the recruitment and headhunter front, bring this up in a project meeting at the start of the project. Point it out as a possibility, as the team members are in demand. Indicate that you expect them to contact you before they take any action. Remain vague as to what action you will take.

How to Address the Problem

What do you do when the problem occurs? First, determine what is going on. Why is the person thinking of leaving? Do not plunge in, in an attempt to save the person for the project. This shows desperation and may raise more panic among the team members. Embark on several parallel tasks:

- Find out what the team member's reasons are for leaving.
- Work with the team member to define what must be done before the person leaves.
- Analyze the project plan to see how the project can be restructured to keep momentum.

This information can help you determine whether you must make the effort to retain the person on the team. You may need to meet with the person's line manager to discuss the issue and to plan for a transition. Concentrate your attention on the project and not on the person. Some project managers neglect the project and focus on keeping the person on board. For you as a project leader, the project comes first. In project planning, see if you can reassign tasks and restructure work so that the impact of the loss of the person is minimal. Turnover is an opportunity to shake up the project and team, if that is required.

What do you do with the project team? Contact each person and answer any individual questions. Be honest and tell them why the person is leaving. Indicate what you are going to do as the project leader. Be positive. The person leaving is going for a new opportunity. This would not have been possible without the project. The situation allows the remaining team to be involved in restructuring the work and having a say in determining the type and characteristics of the replacement person. If you sense other people are thinking of leaving, then hold off on a project meeting. Instead, follow up as was discussed earlier. Prior to the project meeting, make an effort at restructuring the project based on input. Also, start the process of finding a replacement. At the project meeting, review the situation and discuss what is going to happen next.

Issue: Lack of Commitment

Commitment to a project involves attitude and feelings. The benefit of commitment is difficult to quantify. You have seen projects in which the lack of commitment among the project team meant failure of the project. The team members just could not resolve some problem or issue. There is evidence that commitment can spur greater effort in a project at a critical time. Commitment is also affected by time. At the start of the project, people often feel tentatively committed to the project. During the project their level of commitment or dedication can rise or fall. At the end of the project everyone feels committed because it is near the end and they want to see the project completed.

How This Occurs

Lack of commitment can be due to lack of interest. This is most often the case among part-time team members who are involved in many other projects. They do not have the time or dedication to be committed to only one of these projects. Fortunately, less commitment is often required among these team members. Where commitment matters most is among the core team members. These people often have to spend extra hours and reschedule their lives to accommodate the project.

Individuals can lose their sense of commitment for a variety of reasons. Their commitment may have been very high at the start, but they got burned out in the project. The project leader may not have recognized their efforts. Management begins to assume that they can devote 100 percent of their lives to the project. People then get burned out and seek to leave the project. A project that must function under this level of burnout is referred to in some companies as the "death march project." Events in the project can also lead team members to become discouraged.

Potential Impact

Lack of commitment is not felt in the everyday and routine tasks in a systems project. The crisis occurs during integration, final testing, and debugging, when extra effort is required. If one person lacks commitment, the attitude can rub off on others, affecting the entire project. Lack of commitment then translates into lack of effort. The schedule begins to slip.

How to Prevent the Problem

You, as the project leader, must seek to involve the team members in all aspects of the project and project management. By handling issues, actively updating individuals about the project, and reviewing each other's work, involvement increases. Positive involvement and progress naturally lead to commitment.

How to Address the Problem

You may be able to sense a lack of commitment among some team members by their jokes and comments about the project and their work. When you notice this, you first want to determine what underlying problems have led to this situation. What you do depends on the project state and status. If there is no crisis or schedule push, then you have time to get to the root causes. If there is a crisis, then work with each team member to define a work schedule that will accomplish the project goals while at the same time not exhausting the team member.

ISSUE: LACK OF KNOWLEDGE

Project leaders often make the mistake of assuming that certain team members have specific levels of technical knowledge or that they have extensive knowledge of the business process. Then the leader is surprised to find out that the knowledge is lacking. This has become more commonplace due to the wide range of technical and business knowledge that a project requires today. A project manager may expect gaps and holes in the team's knowledge.

How This Occurs

The demands of the project for detailed knowledge have been cited as one reason why lack of knowedge occurs. Another factor is that people naturally do not like to reveal what they do not know. This is especially true for individuals who are hired into a company to work on a specific project. They may overstate the extent of their experience. If the project leader fails to assess the skills and knowledge of the team members in advance, there are likely to be many unpleasant surprises in terms of knowledge gaps.

Potential Impact

Lack of knowledge often does not surface until the project and tasks have been active for some time. Individuals, on their own, may try to hide the gap or work around it. The first impact is the slippage of the task involved. If the knowledge is critical, the project leader may have to find new resources to fill in—adding to both expense and duration. There are also social impacts. The other members of the team may no longer trust the team member. Conflict among the project team members can be created. The person may feel depressed from finally having to admit that he or she lacked the necessary skills.

How to Prevent the Problem

The first step is to carefully define what skills are needed. Most project leaders provide only vague descriptions instead of detailed skill requirements. The next problem occurs when the candidates for the project team are interviewed. The project leader may fail to identify questions that test their knowledge and skills. This can cause the wrong people to be hired.

Another situation is that the manager is so desperate that he or she conducts only a cursory evaluation of skills. The manager just wants to fill up the project team. This occurs often when companies want to hire people with scarce skills that are in demand (for example, people skilled with client-server, extranet, or ERP systems).

Identify team members and others who have the proper technical background to evaluate individuals for the team. Have them develop specific questions and review them with you. These questions are often best if they present a situation to which the person can respond.

An individual who lacks certain knowledge or skills may still be suitable for the project. This is especially true for business unit staff who may not know all of the detailed steps in processing exception transactions. The project leader encourages people to identify gaps or lack of knowledge. Then the two of them can work on identifying alternative solutions to address the issue. This is a more proactive and positive approach.

How to Address the Problem

If a team member lacks the knowledge to do the work, then you first must consider the schedule. Is there time to allow the person to accumulate this knowledge? If there is, then you can send the person to training. If there is not, then you must help identify additional resources that can fill the gap. Your goal is to find someone who can work with the team member and transfer knowledge as the tasks are performed. This will provide guidance to the team member and make the person more self-confident.

ISSUE: TEAM MEMBERS ARE INFLEXIBLE

How This Occurs

Flexibility is a vague word. You can be flexible in the use of a method or tool, but locked into only those methods and tools. Inflexibility is not uncommon in systems. Because people put a great deal of time and effort into learning the technology, they are reluctant to discard it. There is psychological commitment to the method as well. Legacy systems are often sustained by individuals who employ the same tools and methods for years. As a side note, you can easily see that a major barrier to new technology is the speed with which people are willing to adopt and support it. When a technology is new, there is even more reluctance because they don't know if the technology will last.

Potential Impact

Inflexibility may impact the quality of work through the continued use of old methods. It also may mean resistance and disagreements within the project team between groups who favor the old and new. Having multiple approaches can disrupt the project.

Inflexibility may not be obvious; impacts may only appear later. A COBOL programmer, for example, can be trained in C++ and object-oriented methods. Then when he or she starts programming, the code may resemble COBOL and is not object oriented. The impact of inflexibility can be poor quality work in the project. Different individuals may adopt different styles and methods for using the same tools. Thus, the project may finish, but maintenance may be a nightmare. A more severe impact may be that your seasoned legacy system programmers resist learning client-server systems—resulting in staffing problems.

How to Prevent the Problem

Anticipate that many people will have limited flexibility. With technology change flexibility is difficult to achieve over time. Appeal to an individual's self-interest in learning the technology. Carry out the following steps:

- Bring the people up to date on the technology and explain why it is important for them.
- Review with them why specific products, methods, and tools were selected.
- Indicate how you will support them in learning and using the technology.
- Identify sources for support and additional information that will be available.
- Clearly indicate the goals and expectations that you have for them with respect to the technology and why these are reasonable.
- Specify the benefits that will accrue to them in terms of their own careers.

How to Address the Problem

If you encounter people who are inflexible and who resist the technology, do not attack them head on. Instead, visit them and elicit their concerns. Do not address these fears at this time. Go back to your office and identify steps that you can take to allay their fears and concerns. Now revisit each concern with them along with what can and will be done. A further step is to meet with the team and indicate what is available. This reinforces your support of the technology in the presence of the team.

ISSUE: TEAM MEMBERS RESIST PROJECT MANAGEMENT

How This Occurs

Many creative people have been abused by project leaders and project management. They may have been hounded to death about status every day by a project leader who lacked both managerial skills and technical knowledge. It is not a surprise that they are turned off to you.

Potential Impact

Team members who resist project management may rub off on other team members. It will likely become more difficult to run the project. Dealing with project issues will become more complex.

How to Prevent the Problem

At the start of the project, discuss how the project will be managed. Follow the earlier suggestions and lay out a role for the team members by identifying their own tasks, updating their tasks, and identifying and helping to resolve issues. Follow this up with actions that reinforce this. Establish patterns of behavior early in the project before these are major problems.

How to Address the Problem

If you sense resistance, ask the person privately about past projects. Sympathize with the person and show how your approach is different. Reinforce the methods described earlier. It is possible that the resistance to project management masks other underlying concerns about the person's work.

ISSUE: CONFLICTS WITHIN THE TEAM

A team is composed of individuals working together. Many systems managers attempt to manage by emphasizing individual work over teamwork to minimize conflict. This often leads to other coordination problems related to integration issues.

How This Occurs

How can conflict arise in a team? People can clash regarding which methods or tools to use, how best to employ the technology, personality conflicts, and what constitutes acceptable quality—just to mention a few. The conflict may not be noticeable at first. It may surface over an issue at a critical point in the project. Conflicts can arise between junior and senior staff or between technical and non-technical people.

Potential Impact

Some team conflict is natural, unavoidable, and positive due to various possible technical solutions as well as different experience. If not directed toward resolution, then the problems can affect the schedule and productivity can suffer. Hostility can later resurface and affect decisions in the project.

How to Prevent the Problem

Indicate at the start of the project that conflict is almost inevitable. Resolving the conflicts will be done for the good of the project. There are no winners or losers. When you sense a conflict, make it impartial and deal with the underlying issue.

How to Address the Problem

Try to separate out the symptoms of the conflict, the issue itself, and potential solutions. Depersonalize the conflict as soon as possible. Attempt to address several issues at one time.

ISSUE: TEAM MEMBERS SPEND TOO MUCH TIME ON THE WRONG TASKS

Any individual has work preferences. On the weekend, you probably like to work on some things more than others. It is not surprising that this occurs in systems projects in which people have to work on a variety of widely different tasks.

How This Occurs

Given people's preferences and a lack of leadership, it can be understood why people gravitate to the tasks that they prefer. They feel that they can accomplish something even though it is for tasks of lesser importance. Certain tasks are more interesting than others to them. The project leader may treat all of their tasks as equal, sending out the wrong message.

Potential Impact

If the project leader does not address this issue, the progress on the project can continue, but the actual work on the critical tasks will suffer. The overall project schedule may later slip.

How to Prevent the Problem

To head this off, as the project leader, clearly identify what tasks are most important and why they are important. Cut the team member some slack and indicate that it is okay to work on other tasks as long as critical tasks are addressed. Encourage a multitasking work environment, in which people work on several things at one time. This gives variety and can be narrowed to one critical task and then reopened to several.

How to Address the Problem

Assuming that you are aware that an individual is not working on critical tasks, what do you do? Experience shows that an initial carrot and stick approach can be successful. Indicate how important the critical tasks are, but also allow team members time for other work. Follow this up by visiting team members to monitor what they are doing.

ISSUE: TEAM MEMBERS ARE OVERCOMMITTED TO PROJECTS

There are many projects as well as regular work such as operations support, maintenance, and so on. People today are spread thin. Individuals with critical skills and knowledge are spread even thinner.

How This Occurs

In traditional project management, overcommitment is more likely and the impacts are more severe. Many project leaders are trained to treat all team members as fully committed to their projects. They may then sometimes harass team members to work on their project.

Potential Impact

With multiple commitments a person may feel too much pressure. Then work might suffer. In extreme cases, the person might leave the project.

How to Prevent the Problem

To avert problems associated with overcommitment, become informed of all of the person's commitments. Then you can identify realistic minimal expectations for the time that you require. Monitor the person's time and, when possible, encourage team members to do their other work. If there is a conflict, be reasonable and consider what additional time you can allow.

How to Address the Problem

When there is a problem in the project work, approach the individual and review all of his or her commitments. If necessary, approach other managers that are involved and work out a compromise schedule.

ISSUE: THERE IS A PERSONNEL GAP—MISSING SKILLS

How This Occurs

You thought that you had everything covered, but you find that some people lack certain skills. Or as the project progresses you determine that new skills are required.

Potential Impact

Tasks involving the missing skills slip. Even if you attempt to work around the problem, the impact remains. If it continues unaddressed, the work of other team members will suffer.

How to Prevent the Problem

You can mitigate the problem by assessing the skill levels of the team. Have other team members or outsiders to the project conduct skill interviews. You can also establish and maintain a list of consultants or contractors.

How to Address the Problem

In the event of a gap or hole, consider help from inside and outside the firm. Go over the options with the team members so that they are involved. Have them participate in the evaluation. Plan ahead as to how you are going to bring the new people onto the project with minimal disruption.

Issue: A Team Member Is Reluctant to Leave the Project

How This Occurs

Individuals may establish friendships with other team members. They may feel secure in being on the team. This is a known factor for them as opposed to an unknown. They may be reluctant to move to a new, unknown situation where they have to form new relationships.

Potential Impact

The productivity of the team may suffer as these people hang on. Your credibility as a project manager may suffer along with your budget. The team wonders, "Why isn't something being done?"

How to Prevent the Problem

Clearly identify the end point of each team member's services to the project. In addition, as the work of individual team members winds up, help them reestablish their ties to their home line organization. This will make the transition easier.

How to Address the Problem

If you sense that someone is attempting to hang on, schedule a meeting with the person's manager. Don't hint at problems. Indicate that it is in everyone's interest to establish and carry out a transition.

Issue: Team Members Spend Too Much Time in Communications

How This Occurs

Sometimes a subgroup of the team enjoys socializing too much. These team members take up project meeting time with their personal lives, stories, and so on. In addition, they spend considerable time meeting among themselves. It is natural that this will occur on long projects, in which people may bond through a common experience.

Potential Impact

Although some communications is fine, excessive time may be consumed. In addition, other team members may feel shut out from the project—they sense a clique. Internal team conflicts may expand.

How to Prevent the Problem

Indicate after the start of the project that some communications is essential, but that friendships and other relationships should be carried on outside of the project. Make sure that all team meetings address issues.

How to Address the Problem

It is better to redirect the communications to the project each time personal items are brought up. Reinforce this behavior constantly. You may have to reorganize tasks to head off having any subgroup of team members become a clique.

ISSUE: A TEAM MEMBER RESISTS LEARNING NEW SKILLS

New technology emerges at a continuing pace. Since ancient times people have resisted change. They feel that the skills that they use today are still valid and will continue to be in demand in the future. Changing this attitude is very difficult and becomes more so as people gain more experience and seniority in their positions. The point can be made that this issue must be treated by the line organization in which the person is based. However, it is typically the project that uses the new technology. The line organization only requires the older skills and technology. This can create even more pressure for the person to leave the project or to turn down the opportunity to join the project.

How This Occurs

When the project leader indicates that the team members will be able to learn new techniques or technologies, he or she thinks that this is a major benefit from serving on the project. After all, the team is getting paid to be trained, to gain expertise, and to use the technology. This raises the team members' skill levels and makes them more salable in the market. Sounds good, eh? The argument will work with many on the project team. However, there will be some holdouts. To some, a new technology represents a series of threats.

- Some people may feel that their current knowledge and experience will be worthless. They can begin to feel obsolete.
- They may feel overwhelmed. Perhaps, they have not learned new technology for a number of years.
- They may believe that the existing technology is better than the new technology.
- Business unit staff may feel that their current process is just fine and has served them for decades without new technology.
- Some people prefer stability to change. They resist new technologies in general.

When personal computers first emerged, mainframe programmers were some of the last to adopt this technology. In ancient Rome adoption of new technology virtually halted in the second century A.D. An example was the failure of the Roman Empire to adopt stirrups for horses. Stirrups provide stability, thereby allowing the rider to fire arrows more accurately. The enemies of Rome used stirrups; Roman armies paid the price. The problem in systems is typically most acute among programmers. Some COBOL programmers who learn a new language such as C++ continue to design and code programs with a COBOL mentality.

Potential Impact

It sounds simple. If a person resists, you can replace that person and find someone else. Or you can place pressure on the person. It is much more complex. Resistance may not be open. It may not even be evident until after the person has been trained in the new technology. Unfortunately, the person who is reluctant may often possess critical knowledge about the legacy systems and their business rules. The person cannot be replaced and will be needed. You may now have to hire additional staff.

If a business unit employee resists the new technology, there can be additional problems. The employee may return to the business unit and criticize the new technology. This may create, in turn, problems at the management level. People not familiar with working with technology may feel intimidated. They may feel that they are under too much pressure.

How to Prevent the Problem

As a project manager, anticipate that this will occur. You can take the following steps to pave the way for the new technology:

- Discuss the potential benefit of the technology and what other organizations and companies are doing with it.
- Point out the demands that exist for knowledge of the technology.

- Carefully indicate what improved skills each person will gain.

Next, turn to the learning process.

- Indicate the steps in learning and training and explain that help will be provided.
- Define your expectations for what the people will be able to do in the early stages of the use of the technology.
- Encourage them to voice their questions and concerns.

The item that project managers seem to miss or pass over lightly is that of the expectations. Somehow some project managers think that individual team members can ascertain expectations through osmosis.

How to Address the Problem

When this issue is detected, go directly to the person and solicit his or her concerns. Do not respond defensively about the technology. Show that you are open minded so that the team member will voice all concerns. Also, do not attempt to address these concerns now. Visit other team members and see if they share some of the same concerns. Do not openly ask; instead, get their reaction to how the project is doing in general. See if they then volunteer information.

ISSUE: A TEAM MEMBER LEAVES, PRODUCING A GAP

How This Occurs

Team members can leave a project team regardless of project duration. There can be many causes. They may find work that is more interesting elsewhere. They may not get along with the people on the team. They may not be performing well. Be on top of the situation to sense problems before they are announced.

Potential Impact

Much has been written about the crises that arise when people leave a project. In systems, there may be a gap, but not necessarily a crisis. The damage depends on the expertise and knowledge of the person leaving.

How to Prevent the Problem

If you stick to the approach of just-in-time arrival of team members and their release as soon as possible, the problem can be minimized. For critical team members you must consider backup and cross-training from the start. If you implement this later when there is a problem, you can provoke a crisis.

How to Address the Problem

When someone leaves, view it as an opportunity. Meet with other team members to determine a way to reformulate and reassign work. Then you can identify the remaining gap and fill it in.

Issue: The Quality of a Team Member's Work Is Inadequate

What is quality? Who defines what is acceptable quality? Are team members supposed to assume some standard that is not stated? Quality can mean different things to different people. Let's assume that the work of a team member is not acceptable. It may be that it has to be redone. It may mean that the end product failed. Important here is not only the work, but the impact on the project. If there is a major impact in terms of cost, resources, or schedule, then the project manager will have a different course of action than if there is no effect.

How This Occurs

Poor work quality can result from a misunderstanding of the requirements. The work is measured against a standard different from what the person knew about. Many quality problems result from such miscommunications. They can result from communications problems between team members or from poor individual work.

Potential Impact

At a minimum, poor quality can mean time spent reworking and correcting problems. From an interpersonal view in the project, it can mean that the team members begin to lose confidence in the person's work.

How to Prevent the Problem

At the start of the project, indicate what is expected in terms of quality and how work products will be reviewed. Also, point out what actions are possible in terms of working with the person to improve quality.

The second step is to monitor the work early in the project to establish a pattern. Peer review is a useful way of handling evaluations. During the project watch for signs of potential problems. One common sign is that the work is lagging behind schedule. To make up the time, people sometimes sacrifice completeness and quality. Another possibility is that team members are having a major technical problem and have not made this known to others.

How to Address the Problem

Let's assume that you encounter a quality issue. First, determine the team member's own assessment of his or her work. If the person thinks that he or she is doing a good job, then there may be a training problem. If the person admits concerns, then you both can work to identify what can be done. Volunteer to get assistance for these team members, even if they do not admit problems. If you confront them, you risk making the situation worse.

ISSUE: THERE IS CONFLICT BETWEEN JUNIOR AND SENIOR STAFF

How This Occurs

Senior IT staff members often have the knowledge and wisdom that come from working with the same software, languages, and tools over many years. Maintenance and support are so important that management does not encourage them to actively learn new methods. Thus, part of the conflict stems from different methods and tools. Junior staff see benefit in the new technology; senior staff may see comfort in the existing technology.

Potential Impact

The potential impact is serious. It can lead to open conflict and an unwillingness to work together to resolve issues or perform tasks. In some instances we observed, the work was delayed by over a month due to this conflict.

How to Prevent the Problem

You must assume that there will be conflict. This especially occurs with newer staff in e-business situations who may view older programmers as cavemen. To the older programmers the younger ones appear like pseudo programmers. To prevent the problem, acknowledge that it might happen. Encourage joint tasks from the start to reduce the problems later.

How to Address the Problem

When the problem arises, our approach is to tackle it head on. That is, we would focus on interfaces and integration to get the people to work together. Point out that these joint tasks, while unpleasant, are essential to the project. Indicate that team members can return to their separate tasks later. Actively monitor these joint tasks.

ISSUE: THE NEW TEAM MEMBER DOES NOT FIT

How This Occurs

In any team, a new member might not fit in. This occurs in both business units and IT. In business units, the lack of fit may be caused by personality or procedures and policies. In IT it can involve methods and tools as well as personality. In IT the problems tend to be more striking since the individuals have a closer relationship to methods and tools. A number of IT staff members also may have strong technical skills, but they may lack personal and behavioral skills.

Potential Impact

One impact is that there is a lack of communications. This can cause problems in work quality as well as schedule. Another problem that may arise is that work slows down for many people because they are disturbed by the unpleasant work environment.

How to Prevent the Problem

The key here is how you recruit and bring on new team members. Beyond the normal interviewing and evaluation of skills, you should assess how they work on solving problems and issues and how they work together in joint tasks. After the person is on board, schedule a project meeting where the new member meets other members of the team and where they can talk about lessons learned from previous projects. These steps can help to prevent the problems as well as provide an early warning system that such problems exist.

How to Address the Problem

If you find that the new team member is not fitting in, then you should not ignore it. Instead, you must face it. After talking with individual team members to discover the source of the issue, you can work on the issue and take some actions.

Potential useful actions are to encourage and lay out joint tasks. This will support a behavioral change early in the project.

WHAT TO DO NEXT

1. The following table can serve as a checklist for you to use in the project for detecting the symptoms of problems.

Symptom	Date	Incident	Description	Impact

2. The following summary table can be created manually or from the issues database.

Human Resource Issues

Issue	Priority	Date, age	Assigned to	Decision

Chapter 15

Management Issues

INTRODUCTION

In the past management gave less attention to IT projects, which were treated as peripheral to the business. In the past decade the situation changed as management realized that systems and technology could not only reduce costs, but could also provide competitive advantage and increase revenue. Process improvement and e-business are two factors that have further increased management interest and involvement in IT. While this was greeted by some as favorable, it has been a mixed blessing. Many managers lack knowledge of IT and the time and complexity to implement new systems and technology. Vendor promises, advertisements by manufacturers, and success stories have fueled both involvement and problems. A basic point here is that experience shows that trying to teach IT to many traditional managers is not always productive. A better approach is to work with managers on IT-related issues. Managers are used to solving problems, so addressing issues in a business way is useful.

ISSUES

Issue: Management Changes Direction of the Project

How This Occurs

Due to internal business needs, changes in departments, shifting priorities, and other reasons, management can change the direction of the project. One possibility is that management did not understand the project at the start. This is especially true if there was no project concept defined so that additional ideas later might create more changes and problems.

Potential Impact

The impact on the project team is to slow down or halt work while people attempt to determine what to do in response. While a project can withstand one or two of these impacts, any more than that and people on the team begin not to take the project seriously.

How to Prevent the Problem

The best prevention is to recognize that management's direction might change and to address this possibility at the start, when the project concept is first defined and discussed. In fact, most of the issues in this chapter can be added to your initial list of issues.

How to Address the Problem

When changes occur, go to the business unit managers who are involved in the project and attempt to assess what the change means. Do not make changes until both you and the business manager discuss them with upper management. Otherwise, there could be misinterpretation.

Issue: Management Loses Interest

How This Occurs

Some managers show interest in a project at the start. Then as time passes, they lose interest. This can happen for a variety of reasons. First, the project may have been a diversion, and once they have returned to their other work they really don't have time to deal with it anymore. A second reason is that you have not kept them informed so they just assumed that things were okay. A third reason is that they sense that the project is in trouble and they don't want to get involved.

Potential Impact

The impact may at first seem to be negative, but loss of interest will let you and the team get down to work without as much interference. The real loss comes if you need to have an issue addressed that requires managers' support and you are unable to get to see them. Issues then take longer to resolve and the solution may not be to your liking.

How to Prevent the Problem

Keep management informed from the beginning. However, you do not want to run to management with many issues. If you do, then the managers may perceive that the project is in trouble. Focus on informal communications.

How to Address the Problem

You will become aware that managers have lost interest if you find that they do not have time for you or if they do not express any concern. Keep them updated and surface an issue every now and then to keep them involved.

ISSUE: KEY MANAGER WHO SUPPORTED THE PROJECT LEAVES

How This Occurs

There are several versions of this issue. One is that the manager leaves the company. Another is that he or she is transferred. A third is that the manager loses power and so can no longer support the project.

Potential Impact

If the project does not have a champion, it will be difficult to find people to take issues to. Other managers may associate the project with that person and may not see any self-interest or organization interest in the project.

How to Prevent the Problem

The solution is to have more than one champion on the project. At the start, indicate to them that it is in their own best interests to have several managers supporting the project. You are in essence building an informal steering committee.

How to Address the Problem

Contact the manager and get ideas for who could be a replacement. Also, contact the business unit manager. After all, he or she is the one who is most involved in the project and who will receive the benefits of it. The business unit manager should be the one who finds support for the project.

ISSUE: MANAGEMENT EXPANDS SCOPE

How This Occurs

Like other issues in this chapter, this can be due to external or internal factors. If the project was not defined during its conception, then the scope of the project is not clear to all of the managers. Each person may have a different idea of what the project is to accomplish. This is a common cause for scope creep.

Potential Impact

Expanding the scope is often not accompanied by either more money, people, or time. Whatever you deliver then will possibly be unacceptable. The project may be viewed as a failure even if it did achieve the initial goals within the original scope.

How to Prevent the Problem

During the project concept, hammer away at getting a common view of the scope of the project. Make scope creep an issue. Indicate what factors are outside of the scope of the project. Also, show what happens if there is scope creep.

How to Address the Problem

One way to address scope creep is to review the entire project concept again. Meet with managers and business units to redefine and review the scope. Another alternative is to see if you can divide the project into phases so that the added scope items are included in a later phase.

ISSUE: THERE IS NO WILL TO ALLOCATE RESOURCES

How This Occurs

You received approval for the project. You think this is great. Later, you find that management is unwilling to go to the IT manager or the business unit manager and redirect resources to the project. There are other priorities. Moreover, if the manager orders the redirection, there may be resistance and complaining about operational needs not being met.

Potential Impact

The impact is that people are assuming that the project is going ahead since it was approved. The project is delayed since you cannot obtain the resources. The schedule slips.

How to Prevent the Problem

To prevent this from happening, raise the resource issue at the start of the project. Work with line managers and IT managers to line up resources. Make an effort to show how the project is in their own best interests. This should all be done before you get formal approval.

How to Address the Problem

If the problem occurs, then you can return to the manager to seek support. This, however, is an extreme step and puts the manager in an awkward position. It is better to work with the line and IT managers to get the resources needed to get the project going. Once you show some initial results, then you can return to these people to request additional resources.

ISSUE: MANAGEMENT RULES BY CONSENSUS

How This Occurs

Here management cannot seem to make decisions at the individual manager level. There is a fear of making decisions without involving other managers. Therefore, no decision is made.

Potential Impact

The direct impact is that decisions on issues or priorities are delayed. The burden falls on the project leader to line up support with different managers. This delays the project and diverts the project leader from dealing with the other project issues and tasks.

How to Prevent the Problem

Prevention of this problem begins with an assessment of management and its style in decision making. If you detect that ruling by concensus may become a problem, then you should bring it up with the business unit managers. Enlist their support in working with senior management.

How to Address the Problem

If the problem occurs, then you should go to the business unit manager and work out a joint approach. You should also make contingency plans for dealing with the situation in the future.

ISSUE: MANAGEMENT LOCKS ONTO HOT TOPICS

How This Occurs

Whatever the topic, management can seem to lock onto a specific project because it involves something that the managers are interested in. It may not even be critical to the business—just something that caught their fancy.

Potential Impact

The result may be that managers are now showing interest in the details of the project. They want to attend meetings and be involved. While attention is nice, it is likely to be counterproductive and get in the way of work. People in the meetings will not talk openly in front of the manager. You then have to hold additional meetings without the manager present.

How to Prevent the Problem

Try to make the project seem mundane while it is important. Play down the new technology and other factors that are attractive to the manager. Instead, you should involve the manager in some issues so that she or he understands more about the project.

How to Address the Problem

If the problem arises, then you might want to discuss the issue generally with the manager. Your approach is to suggest that the manager might want to be involved in issues rather than the details of the project.

ISSUE: MANAGEMENT ADOPTS A SPECIFIC PACKAGE AND JAMS IT DOWN EVERYONE'S THROAT

How This Occurs

This has happened frequently with Enterprise Resource Planning (ERP) systems. Managers at the corporate level decided on the package and everyone had to get in line to install it. This sometimes also happens because of the industry. If everyone is buying a particular package, then there is substantial pressure to buy and install the same package. Vendors apply pressure as well. Managers are then attracted by the lure of the software's benefits.

Potential Impact

The impact can be devastating—not just to IT, but to the business units. Implementing a complex package requires the precious few experienced and knowledgeable users. Taking these people away from their everyday work can wreak havoc on the departments. In IT the impact may cause resources to be diverted from important maintenance and enhancements. Systems staff members are forced to juggle between the demands of the package and their normal work—hurting both.

How to Prevent the Problem

The important thing here is not to head off the acquisition of the software. It may be the best decision. Rather, the emphasis should be on ensuring that everyone understands the effort that will be required to implement the software from both the business unit and IT perspectives.

How to Address the Problem

When you receive the mandate for a software package, quickly identify implementation issues related to interfaces, changes to the business processes, and the time and effort required to set up the software to meet the needs of the business. Prepare a list of issues as well as a plan. Identify situations where staff will have to split their time to support the package. You never want to say no, but you do want people to realize the schedule impacts.

Issue: Management Listens Too Much to Consultants

How This Occurs

Consultants are a regular feature of the corporate landscape. Assume that it is natural that these people will want to establish and build ties with management. After all, they are seeking more work later. They hope to get it by aligning themselves with senior managers. They also are trying to protect their positions. Managers as a result sometimes overly rely on these consultants.

Potential Impact

It is not that the consultants are malicious or trying to cause problems. Often their advice and counsel are misinterpreted. The impact may be the appearance of a change in direction or focus. It can also mean that the project is given much more scrutiny. Either way, it is going to take more effort now to deal with this issue—time that could be spent on the project.

How to Prevent the Problem

One approach is to maintain contact with both management and the consultants. Encourage the consultants to express any of the concerns to you first. Don't try to stop them from contacting managers. This won't work and places you in a defensive position. Instead, try to make them feel part of the team.

How to Address the Problem

When a problem arises, you will probably hear about it from the manager. Often the manager will not credit the source, but will just identify the issue or situation. You should be ready for this possibility and if it occurs, then you will work to provide answers to the manager. You should also keep the consultant informed. Perhaps, the consultant will begin to realize the situation that he or she created. You should also assume that you will be given partial information and that you will have to investigate further.

ISSUE: MANAGEMENT WANTS TO CHANGE E-BUSINESS PRIORITIES

How This Occurs

Managers are besieged with information related to e-business. They hear about new industry trends, new products, and what the competition is doing. This in turn generates requests for changes and for additional work. Examples might be targeting new audiences or stressing new or different products and services.

Potential Impact

Changing minor requirements is one thing. Changing direction, schedules, and other factors in the middle of e-business implementation creates many problems and opportunities. The impact in some companies has been that e-business implementation was delayed months as direction shifted.

How to Prevent the Problem

While the problem cannot be prevented since you do not control management, there are some things that you can do. First, base your work in implementation on business processes rather than on an analysis of customer behavior on the Web or elsewhere. This is more stable. Second, keep in frequent contact with managers. Get them involved in issues. If you do this, then you will likely find that they gain a better understanding of what it takes to get e-business operations up and running.

How to Address the Problem

When changes of direction occur, don't immediately change direction. Instead, try to understand the source of the change and the reasoning and logic behind it. There are many different ways to accommodate changes in direction. Don't pick the most difficult. Another comment is to use this change as an opportunity to determine what other changes there might be later. This change in direction might be the tip of the iceberg.

ISSUE: MARKETING DEFINES NEW E-BUSINESS INITIATIVES WITH MANAGEMENT SUPPORT

How This Occurs

In e-business and even standard business, marketing strategy plays a critical role in determining the direction of the company. Marketing attempts to be creative to beat the competition. In e-business there are many opportunities for creativity in promotions and discounts.

Potential Impact

Handling traditional requests usually has meant tight, but reasonable deadlines. In e-business this is not the case. There is pressure to get the changes made to both the systems and processes so that the new marketing campaign can get going. In the print media there is a substantial lead time for publication. For the Web it is much faster and more immediate.

How to Prevent the Problem

Actually, the problem cannot be entirely prevented. The best approach is to determine the range of potential marketing programs so that you can anticipate potential requests before they occur. This will help in shortening the work later after the request is made. Doing this also helps to structure the marketing organization. However, it is not guaranteed since creative promotions arisie every day.

How to Address the Problem

When a request comes up, you are put into a reactive mode of operation. IT and the business units are placed on the defensive. To head this off, you should make a major effort to keep in regular contact with marketing staff and managers to see what ideas they are working on. Making this an ongoing effort will prevent surprises. When you do receive a request, consider meeting with the marketing department and the business units affected to determine the best approach. Always consider nonsystems solutions and approaches.

WHAT TO DO NEXT

1. Identify management issues associated with your current and past projects. Then address the following questions.
   - What was the state and condition of communications with management—both formal and informal?
   - What was the elapsed time to resolve issues?
   - Did management bring issues to you or did you present and update management on issues?
   - After issues were resolved, what was the relationship between management and the project?
2. Prepare a table of the issues presented here as rows and the projects as columns. In the entry of the table indicate to what extent the issue appeared in the project.
3. Looking back on management issues that you encountered, identify what could have been different.

SUMMARY

Management issues will not go away. History shows that many of the same issues recur repeatedly. Anticipate that these and other issues will arise with the same and new or different managers. Lessons learned from solving management issues and communicating more effectively with management are keys here.

Chapter 16

Technical Issues

INTRODUCTION

When you think of technical issues, you often think of those that relate to programming, hardware, or networks. Technology and systems are much more complex today. There are more interfaces and integration between systems. Staffing resources are limited, creating more competition for scarce technical resources. Technical issues thus become more intricate and complex.

ISSUES

ISSUE: LEGACY SYSTEM SUPPORT IS TOO RESOURCE INTENSIVE

Legacy systems will not disappear soon. They have, in many cases, been customized to the specific business environment and business rules. They have adapted to the business. They can only be replaced through custom development or by adapting the business to the software packages. Yet when new client-server, data warehousing, or intranet systems are developed, they often must interface with the legacy systems.

How This Occurs

Legacy systems typically depend on a few key programmers who have lived with them for many years. They provide maintenance, enhancement, and production support. Each person often has unique knowledge and experience with the systems. These people face a backlog of work as well as demands of continuing, day-to-day support. On top of this, there are demands for interfaces. Efforts to use productivity aids and new tools achieve limited success due to the old technology embedded in the legacy systems. Training new staff in the legacy systems is often not possible due to limited resources and the difficulty in finding and training new staff in a system that lacks documentation.

Potential Impact

Users are often frustrated at their inability to advance their capabilities due to these limitations. It is a catch-22 situation—there are not enough resources, yet additional resources cannot be added. The situation becomes more difficult to improve. Eventually, of course, all legacy systems will be replaced. However, the major question is when resources will be made available to do this.

How to Prevent the Problem

Develop a strategy for front- and back-end systems to connect to the legacy system with modern intranet or client-server systems. This will allow business rules and additional functions to be supplied without modifying (only interfacing) the legacy system. This extends the life of the legacy system. You can add a strategy for replacement of the legacy system, if that is possible. You can also define small changes to reduce support as well as take a harder line on enhancements.

How to Address the Problem

Review the backlog of requests for the legacy system and cut it, if possible. Go through the maintenance and enhancement requests to see if they can be addressed in some other system or eliminated. Give the support requirements and limitations of the legacy systems more visibility in terms of their impact and resource consumption so that management is aware of their cost, restrictiveness, and impact.

ISSUE: THERE IS A LACK OF AVAILABLE TRAINING FOR STAFF

How This Occurs

New technology creates training requirements for the technology itself as well as interfaces. When a new technology first becomes available, training is limited and there is a lack of expertise. Employees must learn the technology through some form of training. However, their schedules, the training schedule, and the timing of projects using the technology all restrict availability.

Potential Impact

If training is conducted too early, knowledge can be lost before the technology is used. If it is too late, then training is rushed and staff must learn it on their own in the midst of the project. Lack of training will slow the project and likely create rework later.

How to Prevent the Problem

When evaluating any new technology, determine your training requirements and schedule as well as the availability and quality of training. Schedule training as part of the project plan. Try to find classes that offer tips and lessons learned as well as instructions.

How to Address the Problem

If training is not sufficient, then consider having a consultant do in-house training and provide experience and knowledge in the project to shorten the learning curve. However, ensure that you establish a cutoff time to make the staff self-sufficient.

ISSUE: THE TECHNOLOGY REQUIRES A LEARNING CURVE THAT IS TOO LONG

How This Occurs

To employ new technology, such as that related to software or networks, you must gain understanding and then advance to some level of proficiency. One of the reasons for the failure of early client-server projects was that there was an unexpectedly long learning curve for dealing with the client, server, and middleware. Learning a technology also may require that you unlearn old habits that proved useful in the past. This is particularly true in object-oriented programming versus traditional programming. Management may also divert people from learning the new technology by encouraging them to maintain the old technology. This shows the staff that management places a greater value on the existing technology than on the new technology.

Potential Impact

If you underestimate the learning curve and do not allow time for people to gain proficiency, the project will likely slip. The new technology typically must be supported by training and education, consulting, and support. Additional software tools may be required as well. If these are not provided on a timely basis, the project schedule suffers. Not only can the schedule suffer, but the quality of work can suffer as well. This can occur if people try to gain their knowledge through work on the project—often resulting in substantial rework.

How to Prevent the Problem

Consult with the vendors and existing users of the technology to gain insight into what ingredients are needed for successful implementation and use of the technology along with the estimated time for the learning curve. Develop an approach for measuring how people are doing toward gaining expertise.

How to Address the Problem

If this problem arises, divide the tasks in the schedule and attempt to spread the learning curve among multiple staff members. In addition, work with employees to determine if any other assistance might be beneficial. Hold meetings to share lessons learned related to the technology.

Issue: There Is a Lack of Experience and Knowledge of the Technology

How This Occurs

If the technology is very new, available experience will be lacking. There are likely to be bugs, missing features, and other problems. Alternatively, the technology might have been out in the market for some time, but has attracted few users. Why does anyone want this technology? One reason is that people are desperate to fix problems so they want to use it right away. Another reason is that the technology fills an existing gap in the architecture. Perhaps no competing product is on the market.

Potential Impact

One obvious risk is that you are going to be a pioneer in using the technology—that is, the project that depends on this technology will move from one problem or issue to another. It will be more time consuming to get around problems because the vendor will have to be contacted more frequently. The situation can also occur when you are the first to integrate a particular combination of technologies.

How to Prevent the Problem

Make sure that learning the technology is planned within the project. Gather lessons learned from the vendor and any other users. Involve the vendor staff in the project if possible. You also must identify and pay much more attention to areas of uncertainty involving the technology.

How to Address the Problem

If you begin to run into a succession of problems, consider stopping to gain more knowledge of the technology. Pressure the vendor for additional assistance. Slow the project down and give higher priority to training and learning.

Issue: The Technology Does Not Work

How This Occurs

Not working can mean many things. The technology may not work at all, it may not work the way you require it to, or it may fail due to the specific configuration—such as it was undersized. Whatever the cause, the technology is not performing as desired. This may occur due to underestimating the requirements, misunderstanding of the technology, and vendor error.

Potential Impact

You are fortunate if this issue can be resolved by throwing money at the problem and increasing capacity or capabilities. Examples are adding memory, disks, or upgrading the CPU to increase performance. However, if you are dealing with critical software, then you probably have a real problem. In the worst case, the software may actually have to be dropped and a substitute found. If you retain the software, you may have to work around the technology.

How to Prevent the Problem

Identify the minimal acceptable working level for the technology. Get the vendor to commit to this level. In some cases, you can prevent the problem by putting in additional capacity at the start. You can also attempt to test the technology prior to acquisition or right after installation. However, the problem may surface in interfaces. Thus, you must test interfaces as soon as possible. Concentrate resources there.

How to Address the Problem

If you encounter an interface problem when the technology is failing, consider creating a new project plan that includes only the interfaces. Manage it with much more attention. Involve vendors whose products are used in the interfaces.

ISSUE: ORDERING AND DELIVERY OF THE TECHNOLOGY ARE DELAYED

How This Occurs

Delays can be due to your internal purchasing process, the vendor not having the product ready, or politics or geography, such as when the technology is shipped internationally and is beset by government red tape. Many project managers do not consider the purchasing and acquisition process. They discover too late that they should have begun on this at start of the project. Delays due to problems in negotiations can also occur. People can take the mundane process of ordering and delivery for granted. Some technology projects may require specific manufacturing or customization with substantial lead times.

Potential Impact

The technology is not available. The schedule is at first not impacted. People work on other tasks. Time passes. Then the problem becomes more acute. The direct impact of slippage occurs. Morale might be affected if people begin to feel that if the project were important, the components or technology would be on hand. In one manufacturing situation test equipment was delayed, seriously affecting production.

How to Prevent the Problem

As a project leader, set aside time at the start of the project to concentrate on the procurement process. Understand how this process works. Volunteer to work with people in purchasing to expedite the tasks. Remember that if you show how urgent it is to acquire equipment on time, your effort will likely be met by a similar effort. Follow up in negotiation, shipment, delivery, and verification. Work with the vendors to expedite their efforts.

How to Address the Problem

If you are faced with delays, do two things. First, devote the time to follow up on problems. Second, contact the vendor and see if a duplicate component can be shipped immediately. You can later return the faulty component you first received. If there are purchasing delays (approval cycles, freezes, and so on), then show the impact of the delay to management in tangible terms within the schedule.

ISSUE: A NEW VERSION OF THE TECHNOLOGY WILL BE AVAILABLE SOON

How This Occurs

Technology changes quickly. New releases of some products occur at three-month intervals. For some software, these include fixes of known problems as well as new features, functions, and performance improvements. The situation is compounded because the typical systems organization faces the issue of upgrading across multiple products from a range of different sources. The basic question is, "When should you move to the new version?" The more general question is, "What is your general strategy for upgrading many different products?"

Potential Impact

New versions can create problems with interfaces and integration. Additional errors may surface. New versions can have latent errors that were not detected during testing. If you jump too soon, you risk becoming a pioneer. If you wait, you might miss important benefits. If you wait too long, you can risk obsolescence. You may face additional complexity if you made changes when you installed the last version and did not keep a record of what changes were made.

How to Prevent the Problem

Begin with an overall strategy for upgrading the technology. Which components link together and so must be upgraded together? Where do the most problems lie? What are your problems and how do these impact priorities? Consider creating a formal upgrading project plan.

How to Address the Problem

If you adopt new versions on an ad hoc, unplanned basis, then you can face almost nonstop disruption—like having the street in the front of your house torn up all of the time. If you encounter several potential upgrades at one time, create a small group of systems staff to consider what to do in terms of a strategy.

Issue: There Is a Lack of Support for Interfaces to Existing Technology

How This Occurs

With increased networking and integration among products from a variety of vendors, it is not a surprise that many problems exist in the areas of interfaces. This is complex because you and your staff must learn the technology on both sides of the interface as well as how to glue it together effectively. The vendors will likely only have limited knowledge of each other's technology products.

Potential Impact

If you fail to give interfaces a high priority, you risk the schedule and plan. Everything can be complete except for one critical interface. Without it the project can fail or at least be put on hold.

How to Prevent the Problem

Interfaces are so important that you must plan the project around the interfaces and integration effort. Establish tasks that address interface testing. Ensure that test data and suites are prepared that thoroughly test the interfaces.

How to Address the Problem

If you encounter an interface problem that is likely to affect the project, consider setting up a subplan for the interface. Make yourself the project leader and devote your time to coordinating the tasks. Coordination of the interfaces is both complex and time consuming.

Issue: There Is a Gap in the Technology

How This Occurs

Each time a new technology is implemented, gaps are created between it and the existing technologies. This even occurs with standard interfaces because the new technology adds content or performance to interfaces. For example, a new network operating system creates gaps in network management. Features found in network management software may be included in the operating system. Information formats may be different. Many products are affected by a steady evolution of the overall architecture as new versions of current technology as well as new technology are added.

Potential Impact

What happens if a gap is not addressed? You may not be able to use the technology at all. It is more often the case that you will not be able to use new capabilities until the gaps are addressed. The staff must also learn the technical details of the new technology in order to deal with the gaps. This takes valuable resources away from other work.

How to Prevent the Problem

Identify existing gaps in your current technology architecture. Develop a strategy for dealing with these first. Next, include the gap issue in the evaluation and selection of new technology. When people suggest new technology, make sure they are aware of the potential problems.

How to Address the Problem

When you encounter a gap, have staff understand the details of what has to be done. Then contact the vendors to see what help you can get. Make gap analysis and contingency planning part of the projects involving new technology. In addressing the gap, make sure that you establish many small milestones so that you can track progress.

ISSUE: THE PILOT RESULTS FROM THE TECHNOLOGY WERE NOT SUCCESSFUL

How This Occurs

Everyone expects success in pilot projects using new technology. After all, newer is better, right? Why do pilot projects fail? There are many possible factors:

- The objectives and scope of the pilot project were not clearly identified.
- The pilot did not address specific technical issues.
- Resources to perform and evaluate the pilot test were lacking.
- Time to do the pilot was insufficient, so the pilot was rushed.

Thus, success hinges on much more than the technology itself. A major airline conducted a pilot project for a new airplane restroom. It was placed in the middle of an open floor of offices. No one used the pilot. However, because there were no measurements or complaints, the airline assumed that the pilot was successful. When the unit was installed in planes, people got stuck in the toilet and the unit had to be removed—a pilot project with a flush?

Potential Impact

What do you do after the pilot? How do you interpret the results? Can you proceed with the project, should you run another pilot, or should you stop the project? This dilemma can hang up a project for some time. If you attempt to fix the problems in the pilot during finishing of the system, then you risk having additional problems surface.

How to Prevent the Problem

A pilot project has specific goals and scope. The process or system must be measured prior to the pilot as well as after the pilot. Criteria for success of the pilot are established at the start of the pilot. These criteria cannot be weakened after the pilot to force acceptance unless all are in agreement.

How to Address the Problem

If you are faced with problems from the pilot, consider the following alternative courses of action:

- Continue with the project and make adjustments later.
- Change the purpose and scope of the project to fit the results of the pilot project.
- Rerun the pilot based on revised assumptions and setting.
- Stop the project and return to the previous stage in the project.
- Substitute other technology and either continue or run a new pilot.

To evaluate these alternatives, you must review the costs and benefits of the new project.

Issue: The Performance of the Technology Is Not Satisfactory

How This Occurs

In a broad sense, technology performance can involve workload, response time, security, reliability, throughput, utilization, productivity, cost, benefits, and other measures. The values of these measures depend on the work submitted to and performed by the systems and technology. If the actual workload is substantially higher than what was estimated, then performance problems are likely to arise. Performance may also not be acceptable because business units or information systems raised the standards of performance. Even with the increased performance of new technology, there are still performance issues because people expect more out of the technology than ever before.

Potential Impact

Performance may not be satisfactory, but it might be minimally acceptable. In some cases, you can continue using the technology while the problem is being fixed. If performance is totally unacceptable, then it may be necessary to halt use of the technology and have the business unit fall back to anther system. This happened in the early days of point-of-sale systems.

How to Prevent the Problem

Ideally, you should test the technology to find the limits of performance and determine at what point problems occur. Also, perform analysis to determine what upgrades are possible when performance suffers. In fact, for every new critical technology that involves potential performance problems, it is a good idea to define a migration path for improvement.

How to Address the Problem

What is the source of the performance problem? Can it be addressed through upgrades? If so, who is liable for the cost? Will the vendor contribute to the cost? If the problem is fixed, will other performance problems surface? You may face a series of problems. Develop a plan for addressing the problem. Make sure that it includes measurement and that adequate resources are applied.

ISSUE: FEATURES ARE MISSING

How This Occurs

Consider software. You carefully evaluated the software package or tool. You verified features by interviewing the vendor. You received demonstrations. However, upon delivery you find that some of the features are missing or are only partially implemented. The source of this problem can be either the vendor or the customer who misinterpreted what was said. You also might have assumed that a feature was present (to support an interface, for example) even though it was not.

Potential Impact

The project can be delayed or at least may have to be changed to adapt to the situation. The question of impact revolves around which features are missing and how dependent the project is on these features. In some cases for software packages, the system may not be able to handle your unique financial, accounting, or manufacturing situation.

How to Prevent the Problem

How do you verify that the technology will do what you desire? The answer is obviously by testing the work through the system. However, setting up for such testing can be very expensive. For software it may involve establishing many different tables. This requires extensive analysis, training, and even more testing. You can talk to references provided by the vendor.

How to Address the Problem

If the missing features are important, you may wish to consider changing the business process to fit the technology as it stands. You can wait until the features are added, but that might take years. You can also implement a procedure for working around the problems.

ISSUE: THE TECHNOLOGY IN USE IS OBSOLETE

How This Occurs

Every business and household contains technology that is obsolete. For example, new technology may have come along that could have replaced what you have. Why didn't you replace the technology? Because your current technology still works the way you want. You don't see any reason to replace it. Another reason is that it is not very important to you and is not used much. You may also determine that the new technology involves more problems and hassle than replacement is worth.

Potential Impact

For stand-alone technology, this is less of an issue. You are probably more likely to continue to use it and avoid replacement. The problems crop up in technology that requires support or is integrated with other technology. For example, let's suppose that you are using an older network operating system that has served you well. Now you want to purchase a software package. However, the package does not run on the current network operating system. In order to use the new software package, you may be forced to upgrade hardware, replace the network operating system, and train or hire staff.

How to Prevent the Problem

Dealing with obsolescence is one of the reasons to define a systems and technology architecture. The architecture can provide you with the information about integration and aging of the technology. The second step is to devise a replacement strategy. Such a strategy identifies the following:

- Which components are to be replaced?
- How they might be replaced?
- What trigger events are required to initiate the replacement?

How to Address the Problem

You face the need to replace the obsolescent technology. This is a good time to review all of the technology and plan an overall replacement strategy. Management will be more supportive of an overall plan that addresses obsolescence as a whole. When marketing and selling the changes to obtain funding, focus on the downside—that is, address what will happen if there is no replacement. Here are some potential impacts:

- The support effort will be increased.
- The firm will be prohibited from implementing new systems.
- The business units will be denied systems that might have major potential benefits.
- The likelihood of failure will be higher.
- It will be more difficult to obtain support.

ISSUE: THE TECHNOLOGY IS NOT SCALABLE TO HANDLE THE WORKLOAD

How This Occurs

Undersizing of hardware and network components for e-business is fairly common. Neither management nor IT anticipated the workload imposed by Web traffic. Moreover, they did not anticipate the high peak load and the fluctuation in the workload. Another reason is that inappropriate components were selected to save money or time.

Potential Impact

The impact can be slow response time or even a breakdown. This situation then leaves to customer or supplier complaints for e-business traffic.

How to Prevent the Problem

Prevention of the problem comes from front-end analysis and gathering lessons learned from similar organizations to determine what their range of workload was.

How to Address the Problem

If this problem occurs, attention must be given not only to fixing it now, but also developing a long-term approach that involves measurement, capacity planning, and acquiring and using systems software tools.

ISSUE: THE WRONG DIRECTION IN TECHNOLOGY WAS TAKEN

How This Occurs

Often the wrong direction is taken because of the past. The most comfortable approach is to just use upgraded products from your current vendors. These tend to be backward compatible so that they are easier to install and use. There is much less training. If you take this approach one time, it is easier each time there is a decision. The end result can be that you have taken a tangent off of the main technology track.

Potential Impact

The potential effect is that you may be stuck with obsolete or inadequate technology. This will mean that you will likely have more turnover of staff as people leave to work on mainstream, up-to-date technology. It also happens that maintenance and operations support will be more time consuming and difficult.

How to Prevent the Problem

One approach to prevention is to assess technology on a regular basis. In that way, you will at least know where you stand. Another approach is to give attention and priority to technology-based issues. This can keep the heat and focus on fixing these problems as well as showing in graphic detail the problems and impacts that the current technology has.

How to Address the Problem

If you find that there are significant technology issues and problems, do not just deal with these one at a time. Instead, you should consider sitting back and defining what a more suitable technology architecture might be. Then you can build on this as a road map for the future.

WHAT TO DO NEXT

1. If you have not done so already, develop a list of technology issues associated with the current architecture. These include the gaps, problems, and opportunities that exist. This is the first step toward improvements.
2. Again, consider the architecture from the standpoint of obsolescence. Define a new architecture that is desirable. Identify both the benefits and the problems addressed by the new architecture.
3. Review your vendor relations in terms of the support provided and the quality of the products and services. Consider implementing a regular measurement process of the technology performance.

SUMMARY

Technical issues involve requirements, performance, support, vendor products, integration, and interfaces. Today, the situation is too complex and the business is too dependent on the technology to risk taking a reactive, ad hoc approach. You must be proactive not only in addressing problems, but also in defining the future direction.

Chapter 17

Vendor and Consultant Issues

INTRODUCTION

Dependence on vendors, contractors, and consultants is not new in systems projects. It occurred when there was a limited number of programmers for a small number of mainframes in the 1960s. Also, many firms outsourced all of their computing to service bureaus. Times have changed, as have the nature and extent of the dependence.

- There is a greater reliance on vendor-supplied application software.
- The use of software tools has increased. These tools are supplied by different vendors but used by the same people.
- Specific systems activities have been outsourced, making the company more dependent because the outsourcer has much of the knowledge of the systems.

In fact, if you consider your own company, you would probably find that it is dependent on multiple vendors for key systems and business processes. With downsizing and limited resources, it is both infeasible and impractical to consider having all of these functions performed by internal support. The goal must be to better manage, coordinate, and direct external support. This will be the theme in the issues addressed in this chapter.

ISSUES

ISSUE: THERE IS A LACK OF SUPPORT FROM THE VENDOR

Lack of support here means that the vendor is not providing sufficient resources to handle the work or that the resources provided cannot address the requirements. This can occur during normal or crisis times.

How This Occurs

The vendor may take you for granted as a customer. The vendor may have supplied support in previous cases, but was not adequately compensated in the vendor's opinion. There may be no formal understanding related to support or how the support will be coordinated. It can also occur that the vendor's staff is stretched thin in that it does not include many senior technical people. The vendor may be concentrating its top people on getting new work and generating proposals. For technology you may be using a version of the product that is being phased out, limiting support.

Potential Impact

Minor problems can fester for some time and lead to a major problem. This has happened with wide area networks, applications software, and operations support. Typically, many facets of the problems with the vendor surface as the situation deteriorates. The effects can spill over to other projects and other vendors. Relations with business units may suffer. People may question the vendor's product and even want to seek a substitute. If not resolved quickly, both you and the vendor may paint yourselves into opposing corners.

How to Prevent the Problem

It seems obvious that the problem might be prevented by defining support requirements and the rules for managing support in the contract. However, people come and go and the ones who may have established such rules on both sides may be gone—long gone. It is important to review and reinforce the rules and guidelines for support each time there is a change, as well as periodically. Important here is the escalation process for problems.

How to Address the Problem

If you detect a pronounced lack of support, sit down and determine when the problem worsened and how it worsened. Did your staff hold up its end in terms of providing information and so on? What events and factors led to the current situation? Go back to the vendor's management and work on the rules for support as well as the current incident. The steps in determining support requirements, providing a vendor with the necessary information, and getting support must be put into procedures. You will also want to institute a logging process for all incidents and problems, if you do not have one already.

ISSUE: THE VENDOR ATTEMPTS
TO TAKE OVER THE PROJECT

This phenomenon is most common in consulting and development projects wherein the vendor sees an opportunity to expand the business and relationship by filling any gaps in project management. This is not all bad, if managed properly, because you do want such vendors to jointly manage the project with you. Also, you may not have the people to address additional tasks that arise.

How This Occurs

Frequently, the situation arises as additional, new tasks surface that require near-term attention. The vendor may volunteer to take on these tasks. If you do not resist, then a pattern may be established. From lessons learned, the most frequent cause of the problem is the lack of a defined role for the vendor and insufficient monitoring of the boundaries between the vendor and the internal staff.

Potential Impact

In some cases, having the vendor take over some tasks can result in excellent work. You are in essence contracting out the additional work as well as some control of the project. A basic question is, "Who is liable if something goes wrong?" This question must be answered early before problems surface.

A more difficult situation occurs when the vendor takes over the project and guides the selection of technology and consultants. This is why it is difficult for hardware technology vendors to be project managers. The impact of an expanded role for the consultant can demoralize the internal staff. They may even begin to wonder who they are working for.

How to Prevent the Problem

Clear definitions of roles of the vendors as well as the process for handling new work are required at the start of the project. Who is in charge must be spelled out. In addition, procedures for handling new situations must be established before the vendor assumes responsibility. These procedures must be reinforced by the internal project manager.

How to Address the Problem

The internal staff and project manager are the first to detect any effort at a takeover. Try to handle this at the project leader level. Look for the cause in terms of a gap in management. Then you can escalate it upward in the organization. However, if not addressed, upper management should become involved.

Issue: The Vendor Delivers Something Different from What Was Promised

This is the case with an application development effort. It can also happen with systems or utility software where features delivered fail to match promises.

How This Occurs

This issue can have several sources. The vendor staff doing the programming are removed from the people who are gathering requirements and specifications. Miscommunications results—you lose. Technical specifications can be misinterpreted. In addition, there can be several alternative means of implementing the same feature—some of which you don't like. What was promised? This may not be clear. Requirements may have changed, but the contract was not updated.

Potential Impact

It is easy to say that you can reject the software or product and send it back. That may delay the project. If what you do receive is workable, you will be tempted to use it if the vendor promises fixes. That may entail developing workarounds (software or procedures that get around the missing feature). You may be waiting a long time for fixes. In one case, a development editor and debugger lacked features related to testing. Another software tool had to be purchased to cover the testing—raising project cost and lengthening the learning curve. More effort is required to convert work between tools.

How to Prevent the Problem

For development projects try to get through the vendor salespeople and analysts to the programmers. Incorporate as much detail into your specifications as you can. Carry out frequent work reviews. Implement a strenuous testing program. Insist on early delivery of design products for evaluation. Conduct the evaluations as soon as you receive them. This shows the vendor that you are involved.

How to Address the Problem

This will depend on the nature of the project. In general, you are faced with the following questions:

- Should you accept the product and attempt to use it as is?
- What is the minimum that must be fixed or changed for acceptance?
- How do you work around the shortcomings in the interim?

Develop an approach jointly with the vendor for dealing with deliverables prior to problems.

ISSUE: THE VENDOR PROVIDES THE WRONG STAFF

Whether your focus is operations, network support, development, or consulting, the vendor is providing people to your organization to perform work. Even if the people are qualified, there can be personality clashes and other problems. The staff provided may not get along with each other.

How This Occurs

The situation may arise because the company did not implement a review process for the vendor staff. There was an assumption that the company was being provided with highly qualified people. Some consulting firms put in a mixture of senior and junior people. Then they may pull out the senior staff. You are left with the remains. Also, the contractors may not be internally monitored.

Potential Impact

Productivity can suffer along with work quality. If the project leader is not on top of the situation, management can wake up too late and the schedule and budget may suffer. Moreover, the credibility of the vendor is at stake. Costs escalate as well due to delays and potential additions to staff.

How to Prevent the Problem

Requirements and qualifications must be in place and reviewed. There must also be an escalation process for replacing vendor staff along with a regular review process with vendor management. Access to the most highly qualified people must be a condition. This is especially true with application software packages. Monitor who is working on the project.

How to Address the Problem

All work by the vendor staff must be halted by the project leader. The vendor provided the wrong people. The project leader draws up a list of the problems and concerns along with a list of possible solutions and meets with vendor management. Don't resort to the power of the purse—holding money back until you attempt to solve the problem at this level and even with upper management.

ISSUE: THE VENDOR TAKES A DIFFERENT BUSINESS DIRECTION, LEAVING YOU ADRIFT

Vendors are independent companies who set their direction based on their goals and management, their perspective of the marketplace, and competition. There is nothing to prevent them from changing direction. They may, for example, stop supporting different hardware or software environments. They may stop providing interface support for specific networks.

How This Occurs

Although the vendor can be blamed or singled out for criticism, it can also be the fault of your organization. If you fail to keep pace with technology, then you are using workable, but obsolete products. To you this is fine because the products work. For vendors it is a nightmare because they cannot continue to support all interfaces.

Potential Impact

One potential impact is that you have to switch vendors and products to continue operation and growth. A second impact is that you, like the vendor, must also change direction. In either case, there can be impacts across multiple projects. In one instance, a company was forced to quickly change network operating systems. This change was disruptive. Not only was development impacted, but electronic mail and network management software also had to change, requiring extensive training. It was a major upheaval lasting six months.

How to Prevent the Problem

Key to the situation is defining and implementing an architecture that employs mainstream technology. You are less likely to run into vendor problems. Maintain regular contact with vendors to sense their direction and learn about their new products.

How to Address the Problem

If you are caught napping, then you must move to quickly redefine your architecture and technology direction—with and without the vendor. Consider the alternative of continued use versus replacement.

ISSUE: THE VENDOR'S STAFF MEMBERS ARE ABSENT FROM THE PROJECT TOO MUCH

Absence can mean lack of physical presence. It can also mean that the vendor's staff members are there, but they are not working very hard.

How This Occurs

Here are some possible causes:

- You have selected a "hot" software package that is in demand. The vendor's staff members are therefore also in demand and are spread too thin.
- The vendor's staff members are not being properly managed and so are not as productive as they can be.
- The work environment has such pressure that the vendor's staff members are burned out or are absent.
- There may be management conflicts or payment problems.

Potential Impact

Lack of productivity is one impact. Perhaps, more serious is that issues that can be resolved remain festering because of the lack of attention by the vendor's staff. If you were highly dependent on the vendor's staff, then your staff may not be able to fill the gaps.

How to Prevent the Problem

Lay out rules for when and where people work. If you are paying for end products and not time and materials, then insist on intermediate deliverables. If you are paying on a time and materials basis, then you must follow up with frequent reviews.

How to Address the Problem

The project leader is responsible for tracking the vendor's staff members and their work. When a problem occurs, the project leader contacts the vendor's manager and informs internal management. If the project is being managed jointly with the vendor, then both can track the people. The vendor should make up for the absence. Here the project leader again evaluates new people assigned to the project. By being the loudest complainer you are more likely to receive attention in the short run. In the long run you must establish the rules for vendor staff involvement.

ISSUE: THE VENDOR'S WORK IS OF POOR QUALITY

Quality problems can result in failure to function. It can also mean that there are many marginal problems. There is also the question of what constitutes acceptable quality—and if the vendor agrees about this.

How This Occurs

Quality is a relative term that may mean different things to the vendor and customer. Quality standards may not have been spelled out and agreed to at the start. There may also be ambiguity as to what quality applies to.

For example, the vendor's product may not adequately support a network interface. There may be problems with software interfaces. The actual software may have bugs or lack code to process specific transactions. It may not be the conscious fault of the vendor. You may be the pioneer user of the particular combination of technology in the specific environment.

Potential Impact

Quality issues may impact performance, cost, or functionality. The issue is the extent to which quality problems affect the business and project. It may mean that the software is unusable until the problem is repaired.

How to Prevent the Problem

Define what quality means for your setting. This may mean setting testing terms for throughput, response time, range of transactions, and error rates. These should be included in the contract. Implement testing and evaluation as soon as you receive the product.

How to Address the Problem

You must quickly determine the impact of a quality program. Without the vendor you must decide on potential courses of action. Then you can involve the vendors to see what steps they can take. Consider the most severe option of stopping part of the project until the problem is repaired.

ISSUE: THE VENDOR'S SKILLS ARE INSUFFICIENT, PRODUCING A GAP

Skills are not restricted to training in a technology. Skills mean proficiency to cope with problems as well as employing lessons learned to use the technology effectively. Skills also mean that the vendor staff can function as a team.

How This Occurs

The vendor delivers people to work on a project. You assume, based on vendor representations, that these people are qualified. Later you find out that they are junior staff members and lack sufficient experience. The vendor may have misunderstood the project requirements.

Potential Impact

You relied on the vendor staff members and they let you down. Project leaders typically find out about this at the worse time—when they cannot solve a problem. Relations with the vendor can deteriorate quickly. More problems can emerge, and progress on the project will be affected.

How to Prevent the Problem

Whomever the vendor provides must be tested in terms of experience as well as knowledge. The project leader designs an initial task that covers this testing in addition to conducting technical interviews. Areas where the vendor staff lack expertise must be identified.

How to Address the Problem

When a problem or gap appears, contact the vendor's manager to set the situation right. The vendor must not only address the immediate problem, but also come up with a longer-term staffing solution that can be monitored.

Issue: The Vendor's Staff Members Seem to Remain on the Project Forever

Projects in systems and technology can drag out due to additional requirements or slippage. Work can come in peaks and troughs. The vendor's staff members remain through all of this. They may stay on during implementation after their work has been completed.

How This Occurs

Why do vendors remain on the project? Because they are needed. Also, they can stay on because of the desire to continue the relationship with the customer. The project may have been poorly planned so that they have substantial downtime or lapses in activity. They may be retained because of their knowledge and as insurance against future problems.

Potential Impact

Costs rise. Internal morale can sink. People begin to ask, "Why are they still here?" Confidence in the project leader can drop. Unnecessary tasks can be created.

How to Prevent the Problem

As the project leader, ask when people can leave the project. This includes both the vendor's staff and internal employees. Identify the final vendor milestone at the start of the project. Don't mention it too often after that, as it can sour the relationship. Keep the vendor focused on near-term tasks, but always remember the final milestone.

How to Address the Problem

If there is a gap in the work or if the vendor's staff members seem to be hanging on, then define final end products and apply the pressure. Begin to phase them out of the project meetings.

ISSUE: THERE IS NO COMMITTED SCHEDULE FROM THE VENDOR

At the start of the project there may be contractual dates for the vendor. These are different from a schedule. The schedule must back up the contractual dates and provide more detail along with interim milestones.

How This Occurs

Committed contractual dates may not be updated due to project changes—affecting the credibility of dates and schedule. The vendor may not have been asked to prepare a plan for work. Or if there was a plan, it was never updated. Also, the vendor's plan may not have been combined or meshed with your plan.

Potential Impact

Problems or disagreements are almost inevitable in larger projects. The vendor's staff is focusing on near-term tasks. No one is watching the overall schedule. There is now substantial slippage.

How to Prevent the Problem

Create an overall schedule that includes the vendor's work at a summary level. Next, have the vendor's manager define detailed tasks to fit under these summary tasks. Review the tasks and their meanings with the vendor. The next step is to have the vendor assign resources, durations, and start and end dates. Review these. Each week have the vendor's staff members update their part of the schedule.

How to Address the Problem

Let's assume that the project has started without a vendor schedule. Don't wait for problems. Insist that the vendor develops a plan to be included in your schedule. Make sure that this plan is updated on a regular basis.

ISSUE: THERE IS SUBSTANTIAL TURNOVER OF VENDOR STAFF ASSIGNED TO THE PROJECT

Just as it can occur within your organization, the vendor can experience turnover of staff. Because the vendor may have no extra or spare resources, this can have a major impact.

How This Occurs

People who work for the vendor may desire more stability and steady work. They may not like the pressure or being moved between organizations and may seek a position with higher pay from a user of the technology. They may be attracted by one of the vendor's competitors.

Potential Impact

When a person leaves a project, you not only lose the set of hands, you also lose that person's mind, experience, knowledge, and expertise. This can create a problem with internal staff. With the vendor's staff there may be warning. The people may just stop showing up or the pace of their work slackens. It has happened. You as the customer may be the last to know. You may be left with vague assurances that the problems will be handled.

How to Prevent the Problem

As project leader, track the mood and temperament of vendor's staff. If you sense dissatisfaction, you have several options. You can consider hiring these workers yourself. However, you may be precluded from doing this by contract. You may want to talk to the vendor's management. However, this can make the situation worse. You may continue to track the problem and work with the person to transfer his or her knowledge to your staff.

How to Address the Problem

Assume that turnover will occur. When it happens unexpectedly, assess the damage internally. Carefully identify what you expect from the vendors. Then contact the vendor and follow up. Strive for as continuous coverage as possible.

ISSUE: THERE ARE DELAYS IN VENDOR RESPONSES TO PROBLEMS, AFFECTING THE PROJECT

There is a problem in the project with the vendor's product. It may require in-depth technical support or a fix. The project can be held up.

How This Occurs

The vendor's staff may be stretched thin. They may feel that your problem is unique and so they assign it a lower priority because it does not affect other customers. They may also lack the expertise to deal with it. The problem may involve products of other vendors. The vendor could have even contracted out that area of the technology. All of these can produce a nonresponse.

Potential Impact

The project team is forced now to work only on tasks that are not affected by the problem. These may not be the most important tasks. The schedule can slip like sand through your fingers.

How to Prevent the Problem

Design a problem-handling process at the start of the project with the vendor. Try to find out how the vendor's internal problem-resolution process works and what methods are used. The more you know, the easier it will be for you to navigate within the vendor's organization.

How to Address the Problem

When a problem occurs, get right on top of it. Don't let up for a minute. Otherwise, the vendor may think that you do not think the problem is important. Establish an escalation process within your organization before a problem surfaces.

ISSUE: THE VENDOR'S STAFF MEMBERS ARE LOCKED INTO THEIR OWN METHODS, WHICH ARE INCOMPATIBLE WITH THE PROJECT

This can occur with planning, reengineering, and other consulting types of work. You thought that you were hiring experience, expertise, and knowledge. To your chagrin you find that you are mainly getting a method and not the experienced staff.

How This Occurs

Vendors often train their staff in their own methods. This has a number of advantages. First, they present a standard front to any problem. Second, the vendor's staff members appear more interchangeable. More important, they believe that the method provides a competitive edge. When you select them for the project, you may fail to understand how they are dependent on their method and how it fits into the project.

Potential Impact

The result can be a lack of fit between the method of the vendor firm and the situation at the customer firm. This can yield continuous conflict. Either the project is restructured around the method, or the method is dropped or modified. The project team members from the customer firm must take time out from the project to learn the method and then educate others as well. The method and consultant may be dropped because the project received a false start.

How to Prevent the Problem

Spend time understanding the method. See how it meshes with your culture and environment. Work with the vendor on the project schedule to see what the impact is of learning the method on the schedule. The method as well as the vendor are evaluated during the proposal stage.

How to Address the Problem

The method turns out to be a problem. It seems that the vendor is force fitting the project situation into the method. The reverse is the right course. Your staff lacks both experience and confidence in the method. What should you do? Consider killing the use of the method before more damage is done.

ISSUE: THE PROJECT IS OVERDEPENDENT ON THE VENDOR

Dependence on the vendor can be for the vendor's products or knowledge and expertise. It is a necessity for software, hardware, and network components.

How This Occurs

You become overdependent when you cannot make project decisions and must rely on the vendor for tactical and strategic direction. This can arise because the internal staff members are not adequately trained in the technology. It can also happen because you outsourced the technical support.

Potential Impact

You can lose control of the schedule and the project. Confusion can reign among your own staff while the vendor calls the shots.

How to Prevent the Problem

Define vendor dependencies as well as the vendor's role at the start of the project. Review these roles with your team. Make the role part of the agreement. Think about a backup approach. Solicit the team's support in monitoring the vendor's work.

How to Address the Problem

If a problem arises, you must meet with the vendor's manager to review roles and responsibilities. You may want to define a plan for transitioning some duties from the vendor to internal staff. You may also wish to involve more than one vendor to obtain backup support.

ISSUE: VENDORS DON'T COOPERATE AMONG THEMSELVES

How This Occurs

More IT projects involve several vendors, consultants, or contractors. ERP, e-business, and process improvement projects are examples. Each vendor typically looks out after its own interests. A vendor may not have estimated or taken into account working with other vendors in the project. Thus, vendors may be reluctant to cooperate with each other.

Potential Impact

The potential problems can slow the project down and create political problems within the team and with management and business units. In some severe cases, the projects have actually been cancelled.

How to Prevent the Problem

Define at the start before the vendors are brought on board the project what you expect in terms of them working together. But this is not enough. You also have to have them work on issues and common tasks together. Try to identify these issues and tasks early and then have the vendors work together. Monitor their work and how they get along. You are attempting to instill a successful pattern of working together.

How to Address the Problem

If the problem arises, you should identify specific issues and instances of problems and then work on these. Don't try to address the problem generally—this method probably will not work because of the generality. By addressing and solving specific issues you can work on building up the cooperative behavior.

ISSUE: VENDOR SUBCONTRACTS OUT THE WORK AND THEN DISAPPEARS

How This Occurs

This occurs when the main vendor lacks the expertise or staff and has to bring in a subcontractor. In such cases, the vendor manager may not set up how the subcontractor is to work with the client firm. Problems begin to mount, and because the vendor is not on the scene, the problems get worse.

Potential Impact

The progress of work can slow down. Subcontractor staff members may make promises and commitments that cannot be honored by the contractor.

Misunderstandings with business unit managers and staff can arise. Sorting it out directly with the subcontractor will probably fail unless you get the vendor involved.

How to Prevent the Problem

Define at the start whether and how any subcontracting will be done. Then define the method for managing and dealing with subcontractors. After the subcontractors are lined up, identify potential problems in meetings with both the vendor and subcontractors. Define how these problems will be addressed. Monitor the initial work of the subcontractors closely to head off problems.

How to Address the Problem

If the problem is detected, then you should immediately involve the main vendor and implement the issues management approach discussed here and in previous chapters.

ISSUE: VENDOR USES WHAT IT DID ON YOUR PROJECT WITH ITS NEXT CLIENT

How This Occurs

This may occur in e-business when the vendor takes the analysis results or code from your project and uses it to gain an advantage over other firms in bidding for the next job. This next job could be with one of your competitors. People have, after all, a natural tendency to employ their experience as they move on to more work.

Potential Impact

If competitive information is involved, this is serious and may call for legal remedies. In less severe situations, you may find that the staff members who worked with you are gone and that when you require additional assistance, you receive resources that lack the experience.

How to Prevent the Problem

To prevent the problem you should first assess the type of work that the vendor is doing and what the risks are to your organization. If there is no potential problem, then you can move on. If there is a problem, you can beef up the contract. A better step is to raise this as an issue with the manager of the vendor firm and determine how it will be detected, prevented, and addressed.

How to Address the Problem

Each situation for this issue is unique. While legal action may be needed, the major leverage that you have over the vendor is future work and references. You can employ these to get some control over the situation.

ISSUE: VENDOR TURNS OVER WORK PRODUCTS THAT ARE NOT USABLE

How This Occurs

The vendor manager on your account may have assigned junior staff members to your project and may not be reviewing their work closely. He or she may be concentrating on marketing to other firms. It may be that their earlier work was not carefully reviewed. Maybe the vendor thought that you really didn't care. Another possible cause is that the client firm has no one on the staff who can technically evaluate the work.

Potential Impact

Unacceptable work can mean later rework. This affects both the schedule and budget. It also creates management problems between the vendor and the client firm.

How to Prevent the Problem

At the very start you should identify what constitutes acceptable work. You should also define an evaluation approach for assessing the work and review it with the vendor. Then when the initial work comes in, you should evaluate it in detail. Another tip is to employ a single consultant if you do not have the appropriate person on board.

How to Address the Problem

Raise the quality of work right away as an issue the first time you notice a problem. In addition, take steps to implement a review process.

WHAT TO DO NEXT

1. For each major vendor, develop a list of issues that you have addressed or that are pending with the vendor. Review your approach for issue

resolution with each vendor. Look for any patterns that indicate where you can improve your approach to dealing with vendors.

2. Assume that critical vendor staff were to be reassigned or become unavailable. What would be your backup plan?

SUMMARY

Outsourcing in information technology can now constitutes 25% to over 50% of the total systems budget. Once you outsource, you lose control while, perhaps, gaining skills, flexibility, and technical support. However, success requires not only careful planning at the start, but also constant monitoring and measurement.

Chapter 18

How to Implement Improved Project Management

INTRODUCTION

A number of ideas and suggestions have been provided in the different chapters. Rather than attempting to use and implement these in a piecemeal way, consider using a more systematic approach. This is the subject of this final chapter. There are three time frames to consider. The first consists of near-term quick hits that can be employed with little effort and hassle. The second consists of intermediate steps that can be taken after the quick hits. These draw upon the quick hits. The third category consists of the longer-term changes that complete the modernization of your project management process. For each we discuss the benefits and give tips on getting it started.

Implementing change of any significant kind is challenging. This can be even more apparent in IT projects that are political. Thus, we have included points of potential resistance to change and suggestions on how to address these concerns.

APPROACH

QUICK HIT IMPLEMENTATION STEPS

- **Identify the projects and nonproject work.**
  You need to identify all of the work that critical staff members are doing in order to allocate their time between regular work and projects. Attention must be given to smaller projects as well. Start with the approved projects and then identify the work performed by each staff member. This will pave the way for general reporting on routine tasks that are not part of a project.

- **Define and use the standardized project binder.**
  The tabs and contents of the project binder were identified earlier in this book. By standardizing on the binder you can retain project files for reuse and lessons learned. The project binder can either be in paper or electronic form.

- **Customize the project management software and adopt standard abbreviations.**
  Customization is necessary in order to be able to use the data in a database type form. Custom data elements for issues, lessons learned, and so forth can then be employed to filter and analyze project data. Without customization there is a lack of standardization and an inability to do multiple project analysis using project management software.

- **Establish a resource pool.**
  A common resource pool is necessary for multiple project analysis just as the customization above is. The pool contains both general and specific resources. In IT projects these resources are people related. However, in other applications you can identify calendar resources that allow you to customize the schedule down to the level of a specific task.

- **Create templates.**
  Recall that the approach to creating templates is to extract high-level tasks and gain consensus on the template. Existing project plans can then be retrofitted into the templates. Templates support both lessons learned and multiple project analysis. Templates also generate collaborative effort and a common understanding of tasks.

- **Identify a checklist of issues.**
  Start with the checklist in the appendix. Then add your own issues by identifying those of the current projects. Use the issues checklist at the start of each project in the project concept. As you find additional issues, add them to the checklist. The checklist reinforces the analysis and emphasis on issues.

- **Modify the approach to project meetings.**
  Several guidelines for project meetings were given. Meetings should focus on either issues or lessons learned. Status should be gathered prior to the meeting and summarized at the start of the meeting. Only team members involved in issues and lessons learned need to attend the meetings. Meetings can be held at a frequency based on the state of the project so that if the project is in trouble, there are more frequent meetings to resolve issues. By adopting these and the other guidelines, your meetings will be more productive and get better results.

- **Standardize the reporting on projects.**
  A one-page project reporting summary was identified along with issues analysis and various tables related to issues, projects, and business processes.

Follow up on these. By using standardized reporting, management can focus on the content of the information rather than spend time trying to understand different formats.

- **Adopt templates and outlines for documentation and presentations.**
  Start with your own projects and develop outlines for all major project documents and presentations with the team. Presentations should include status, budget, project concept, project plan, and issues meetings. Review the previous presentations and documents to see how they fit with the outlines and templates.

- **Set up a model project on the network.**
  Having a project to review and use as a guide is very useful. You should establish a project on the network. The information should include the issues, lessons learned, project plan, and key events in the project. Encourage people to visit the site and provide comments.

- **Gather initial lessons learned from one or two projects.**
  Use the guidelines in the book as ideas for you to gather lessons learned. You will normally identify experience that you gained. Then you will extract the lessons learned. This will provide structure for gathering lessons learned on a routine basis. Also, you will gain experience in establishing the lessons learned database.

- **Develop a project concept from an existing project.**
  Take a current project and identify the business purpose, technical purpose, scope (business and technical), issues, roles, and benefits. Then review this information to see how the project has changed. Pay careful attention to the scope and how it changed over time during the project. This will prepare you to develop the project concept for future projects.

- **Evaluate your contacts with managers and use a forward planning approach.**
  Many of us get caught up in the daily work of projects and normal work. We often do not take time to assess our communications with management. Each week you should review what management contacts you had. You should also plan for contacts in the next week. This will raise your level of awareness and attention for management contacts.

- **Implement the project scorecard.**
  Of the various scorecards presented (project, process, vendor, business unit), the easiest one to start with is the scorecard for your project. On a monthly basis evaluate your project. You can use a spreadsheet in which the rows are the criteria and the columns are the months. You can then assess how you are doing and become more aware of measurement. You can see the forest through the trees.

- **Create a list of bookmarks for Web sites for project management and lessons learned.**
  Begin with the Web sites that have been included in this book. On a regular basis add to these bookmarks. Print out the content and distribute these to your team.

INTERMEDIATE-TERM ACTIONS

- **Using the issues information, create the issues graphs and tables.**
  Having begun to collect issues from projects, you can now construct the graphs and tables developed in the book. This is good material for presentation to management. It is also useful to the project team to see the characteristics and trends of issues. It also paves the way for issues analysis later across projects.

- **Implement the project concept.**
  Having created a project concept out of a current project you can now implement the project concept as a structured approach to review project ideas. Try this out on smaller projects such as network extensions and upgrades, and enhancement requests.

- **Create the lessons learned database and relate it to the tasks in the template.**
  Previously you gathered lessons learned and created a template. Now you can implement the lessons learned database and link these lessons learned to the template. Review the project regularly to see if you can identify more lessons learned.

- **Set up and conduct meetings on lessons learned.**
  You have already modified the approach on meetings and are addressing issues. It is time to raise the level of awareness of lessons learned. Have team members give the results of their experience in team meetings. This should improve morale and facilitate information sharing.

- **Implement a collaborative approach in which team members define and update their own tasks.**
  With the template in place and the changes in project meetings, you are prepared to have team members define their own tasks. Begin with a paper-based approach. Then after everyone is comfortable with this you can move on to updating the tasks by individual team members. Following this you can implement on-line task definition and updating by team members.

- **Scan the literature and collect information on project successes and failures.**

Use the Web to gather lessons learned. We list this as an intermediate task since you will have gained experience in collecting lessons learned. Use magazines such as *Information Week, Ziff-Davis Publications,* and others. Circulate articles to the team members. Encourage them to search for articles as well on the Web and in print media.

- **Employ the process scorecard.**
  Previously you implemented the project scorecard. You can now assess the underlying business processes that your projects are supporting. Do this with team members casually as it will be politically sensitive. This work helps the team focus on tangible benefits of systems and technology and become more aware and focused on the business process. Suggest to management that this be implemented for new projects.

- **Use the vendor scorecard.**
  Another project measurement is the vendor scorecard. Begin with an internal assessment of vendor performance. Share it with the team and then with the vendor. Use the scorecard for new projects.

- **Formalize the approach of bringing new team members onto the project and getting them out of the team.**
  Often new team members are brought onto the project team in a casual way. Implement the structured approach that we have identified. Have them discuss their lessons learned from past projects with the team and assign joint tasks for them with other team members. This will support collaboration.

- **Review the methods and tools used in projects to assess their effectiveness and determine where you have gaps.**
  Create a table of methods and tools. For each area identify the methods, tools, experts to call on, guidelines, and expectations. Also, identify relevant lessons learned. This will support standardization.

LONG-TERM STEPS

- **Begin to address the gaps in methods and tools in projects.**
  Having identified the current methods and tools, you will uncover the gaps. Work with the team and others in IT to fill the holes. By addressing this issue you will not only begin to fill in the holes, but you will encourage people not to invent their own.

- **Review projects and nonproject work and eliminate those tasks that do not contribute to critical business processes.**
  Now that you have identified the work being performed, you can move to eliminate or reduce work that does not add to the performance of the group. This supports focusing on critical activities.

- **Develop process plans to determine the long-term future of processes and use these to identify candidates for new projects.**
  Recall that a process plan is an approach for understanding the future direction of a business process. By developing these process plans you can get the users centered on the longer-term requirements for their business and away from submitting marginal requests.

- **Establish master schedules on the network for all projects.**
  With the templates in place and people updating the schedules, you are in a position to do multiple project analysis and create master schedules. This is a major step toward affecting the strategic direction of IT.

- **Implement a weekly or other regularly scheduled meeting to evaluate resource allocations.**
  With the templates and a collaborative approach, there is a sufficient basis for holding meetings to allocate critical staff resources to project and non-project work on a regular basis. This will not only provide more consistent direction to the staff, but will also yield performance improvements.

- **Begin to involve team members in issues resolution.**
  Issues resolution follows from the collaboration in building and updating the project plan. Team members are now sharing lessons learned. Therefore, it is appropriate to get them involved in analyzing issues.

- **Develop a semiannual or annual report on business processes.**
  Previously you implemented the process scorecard. Drawing on this you can start to analyze business processes on a regular basis. This will help management, the business units, and IT to determine the most beneficial new project opportunities.

- **Implement comparative business unit assessments.**
  This is the scorecard for business unit involvement in projects. Begin doing the assessment and sharing it among IT managers. Then raise the level of awareness among business units for new projects by using the scorecard assessment. This should improve the business unit participation in the projects.

- **Review the list of issues and see if they can be eliminated or handled through lessons learned.**
  Now that you have been gathering and assessing various issues and lessons learned, it is time to consider seeing if some recurring issues can be headed off or prevented using lessons learned.

- **Develop an overall assessment of project effectiveness.**
  Taking into consideration the various scorecards, issues analysis, and lessons learned, you can begin to assess how effective project management and project performance are.

POINTS OF POTENTIAL RESISTANCE AND WHAT TO DO

Some potential areas of resistance to change in general are listed next. While this approach may appear negative, it is not. By reading this material, you become more aware of what arguments you may face and how to address them when they arise.

- **We have no time to learn and employ new methods.**
  People on projects are very busy. Some may feel that they do not have time to consider new project management approaches. However, if they keep doing the same things, there will be no improvement. Point this out. Also, indicate that the methods do not really take more time. Instead, they provide structure for what the team members are already doing.

- **Our situation is unique.**
  This argument is employed to resist a template and standardization. The approach in this book is to apply standardization at the higher levels of project activities and flexibility in the detail. Team members can still be unique in the detail.

- **You cannot force the use of a method or tool.**
  No one wants to force people to do things. It does not work. The methods here support people's self-interest. They provide a framework for managing projects and doing project work.

- **We tried something similar before and it did not work.**
  This may be true. However, you will note that the method is composed of themes that are self-sustaining and supporting. Once you start using the techniques, you will establish successful patterns of behavior.

- **There are not enough resources to implement the changes.**
  The approach can be implemented in small steps without any major change or upheaval. No additional resources are required.

- **Implementing a collaborative approach results in too many interfaces.**
  When you collaborate, you do interface with people. However, there are no more interfaces than what you would face without these methods.

- **There is no benefit in the approach.**
  The approach has been successful in firms in many different industries, countries, and cultural settings. Improved productivity, morale, and performance are the most tangible benefits.

- **The method is too complicated.**
  The approach uses common sense. There is no jargon. It is a step-by-step approach.

- **Change will slow me down.**
  By implementing the approach in small bite-sized parts, no one should be slowed down. With time and experience, team members will become more efficient and effective.

- **The approach reduces the scope for making decisions.**
  When structure is provided, some people may feel that they have less room to make decisions. The approach provides a general framework to address issues.

- **The approach is not how we do things.**
  In general, the method does not change how employees do their work. It provides a framework to support that allows them to do their work better.

- **The method restricts creativity.**
  Creativity lies in the detail not in general activities. Creativity and flexibility are both supported in the templates, issues, and lessons learned.

- **Change in project management is threatening.**
  Any change is threatening to some people. This is understandable. Here we mitigate this resistance by offering a phased method for implementing the approach.

- **What's in it for me?**
  This is a key question and deserves some discussion. The approach frees team members up to do more creative work. With the lessons learned, each team member's work quality and skills should improve. It should take less time to do more routine tasks. Knowing that the same issues have been recurring gives some a feeling of comfort that they are not alone.

SUMMARY

While technology changes and new business initiatives emerge, many characteristics of IT projects remain the same. The goal is to gather and employ lessons learned to reduce the overhead and increase the effectiveness in future systems and technology projects. You have seen how much IT projects differ from regular projects. You have also seen the following themes addressed throughout this book:

- Using collaborative effort in defining, doing, and reviewing work.
- Managing multiple projects and normal, nonproject work.
- Addressing issues across projects.
- Using the templates, issues and lessons learned databases, and other techniques to reduce the workload and make the project leader more effective and efficient.
- Applying lessons learned to improve future performance.
- Encouraging joint ownership of project work by business units, IT, and vendors.
- Tracking and measuring work and incorporating issues analysis.

Bibliography

Angus, Robert B., and Norman A. Gunderson. *Planning, Performing, and Controlling Projects: Principles and Applications.* Prentice-Hall, New York, 1997.

Demarco, Thomas. *The Deadline: A Novel about Project Management.* Dorset House, New York, 1997.

Dinsmore, Paul C. (ed.). *The AMA Handbook of Project Management.* AMACOM, New York, 1993.

Graham, R. J., and R. L. Englund. *Creating an Environment for Successful Projects: The Quest to Manage Project Management.* Jossey-Bass Publishing, New York, 1997.

Knutson, Joan, and Ira Bitz. *Project Management: How to Plan and Manage Successful Projects.* AMACOM, New York, 1991.

Lientz, B. P., and K. P. Rea. *Business Process Improvement.* Harcourt Brace Professional Publishing, San Diego, 1999.

Lientz, B. P., and K. P. Rea. *On Time Technology Implementation.* Academic Press, San Diego, 2000.

Lientz, B. P., and K. P. Rea. *Project Management for the 21st Century,* 2nd ed. Academic Press, San Diego, 1998.

Lientz, B. P., and K. P. Rea. *Strategic Systems Planning and Management.* Harcourt Brace Professional Publishing, San Diego, 1999.

Microsoft Project 4 for Windows: Step by Step Book and Disk. Microsoft Press, Seattle WA, 1994.

Morris, Peter G. W. *The Management of Projects.* American Society of Civil Engineers, Reston VA, 1994.

Randolph, W. Allen, and Barry Z. Posner. *Getting the Job Done: Managing Project Teams and Task Forces for Success.* Prentice-Hall Trade, New York, 1991.

Reiss, Geoff. *Project Management Demystified: Today's Tools and Techniques.* Chapman and Hall, New York, 1995.

Shaughnessy, Haydn. *Collaboration Management.* John Wiley and Sons, New York, 1994.

Spinner, M. P. *Elements of Project Management: Plan, Schedule, and Control.* Prentice-Hall, New York, 1991.

Stevenson, Nancy. *Microsoft Project X Bible.* IDG Books Worldwide, Boston, 1997.

Turner, J. Rodney. *The Handbook of Project-Based Management: Improving the Process for Achieving Strategic Objectives.* McGraw-Hill Publishing, New York, 1993.

Web Sites

PROJECT MANAGEMENT

http://www1.msfc.nasa.gov/TRAINING/APPL/pmdp/handbook/guide.htm
http://www.stir.ac.uk/manorg/images/project_management_sites.htm
http://www.microframe.com/customers/Support/pmlinks.htm
http://www.hq.nasa.gov/office/hqlibrary/ppm/ppmbib.htm
http://jscott.chester.ac.uk/jpgriff/pp.htm   resources for project management
http://www.synapse.net/~loday/PMForum/contents.htm   Project management forum
http://www.hq.nasa.gov/office/hqlibrary/ppm/ppmbib.htm   NASA resource lists
http://www.bized.ac.uk/roads/htdocs/subject-listing/Default/projman.html
http://www.projectnet.co.uk/   project management reference

LESSONS LEARNED

allc.com/_vti_bin/shtml.exe/website/search/searche.htm-250 documents
http://www.em.doe.gov/tie/index.html   Department of Energy site
http://llis.gsfc.nasa.gov/   NASA site
http://www.bmpcoe.org/   best practices
www.stsc.hill.af.mil/crosstalk/1996/apr/project.asp
www.fm.doe.gov/FM-20/lesson/
ara.informatik.gu.se/nuldenHTML/ECIS02.html
www.armyec.sra.com/knowbase/interim/infomgt.htm

COLLABORATION

Set up an intranet   intranets.com
Set up threaded message boards   www.beseen.com, www.server.com
   insert message boards at your web site regardless of server
Set up your own forms   www.response-o-matic.com
Scheduling   hotoffice.com, daytracker.com, webex.com, scheduleonline.com

Appendix 1

The Magic Cross Reference

The letters "ff" indicate the reader should also see the following pages.

Area	Topic	Pages
Develop the project plan	Baseline plan	128–129
	Develop project templates	121
	Develop the project budget	125–128
	Issues databases	116–118
	Issues, lessons learned, templates, and the plans	120
	Lessons learned databases	118
	Methods and tools	116–117
	Risk management	124
	Sell the project plan	129–130
	Task definition	122–123
General	Concepts	3–4
	Differences with regular projects	5
	E-Business and IT projects	17
	Guidelines for success	15–17
	Myths	9–10
	Why IT projects fail	11–14
Multiple projects	Assess current projects	43–44
	Interdependence	41–43
	Project analysis	45–47
	Resource allocation	50–52
	Types of projects	48
Operations, maintenance, enhancement	Departmental systems plans	184
	Measure IT allocation	179–180
	Planning	177–178
	Slate of work	184–186
	Strategic systems plan	183–184
Project concept	Alternative concepts	64–66
	Comparative analysis	66–67
	Content	60–61
	Evaluate the business situation	61–62

Area	Topic	Pages
Project coordination	Budget versus actual analysis	146–147
	Informal communications	148
	Issues management	138–141
	Issues presentations	150–151
	Lessons learned	152–153
	Management presentations	147–148
	Measuring work	141–142
	Milestone evaluation	142
	Multiple projects	146–147
	Project crisis	152
	Project summary	149–150
	Work evaluation	143–144
Project leader	Capabilities	78–80
	Duties	81–84
	Evaluation and selection	80–81
	Managing small and large projects	86–87
	Measurement	92
	Sources of failure	90–91
	Success	91–92
Project management process	Checklist for evaluation	18
	Intermediate improvement steps	320–321
	Long-term improvement steps	321–322
	Marketing	31–32
	Project strategy	21–22
	Quick-hit implementation steps	317–320
	Resistance to change	323–324
	Transition	30–31
Project planning	Management steering committee	25
	Project concept	27–28
	Project scope	24–25
	Project selection	28
Project team	Changing team members	109–110
	Consultants	102–105
	Responsibilities	99
	Skills	99–102
	Team problems	106–108
Software development	Data conversion	170–171
	Design and development	169–170
	Documentation	170
	Methods and tools	158–159
	Requirements	159
	Steps	160–164
Software packages	Benefits and risks	193–194
	Costs/benefits	202
	Critical selection factors	194–195
	Implementation steps	195ff
	Implementation tasks	209
	Software/consultant evaluation	200–201

Area	Topic	Pages
Technology projects	Architecture	216–218
	Costs and benefits	219–220
	Strategy and products	223–226
Trends	Business	6
	Systems	7–8
	Technology	7

Appendix 2
Issues Checklist

Area	Issue	Does issue apply?

Business	The business unit changes requirements frequently
	The business unit does not provide good people for the project
	The business unit is unwilling to change the business process
	Several business units cannot agree among themselves
	The business unit cannot consistently resolve issues
	The business unit staff members do not know the business process
	The existing business process is in poor condition
	Middle-level business unit managers resists change and the system
	The business unit attempts to dominate the project
	The business unit view does not fit with that of upper
	    management or the industry
	Business unit managers and staff lack technical knowledge
	Business unit management is replaced
	The business unit has no interest in the project
	Other work or projects have higher priority for the business unit
	    than your project does
	The business unit has existing technology that conflicts with
	    that of the project	

Human resources	Turnover of project team members Lack of commitment Lack of knowledge Team members are inflexible Team members resist project management Conflicts within the team Team members spend too much time on the wrong tasks Team members are overcommitted to projects There is a personnel gap—missing skills A team member is reluctant to leave the project Team members spend too much time in communications A team member resists learning new skills	

Area	Issue	Does issue apply?
Human resources (*cont'd*)	A team member leaves, producing a gap The quality of a team member's work is inadequate There is conflict between junior and senior staff The new team member does not fit	

Management	Management changes direction of the project
	Management loses interest
	Key manager who supported the project leaves
	Management expands scope
	There is no will to allocate resources
	Management rules by consensus
	Management locks onto hot topics
	Management adopts a specific package and jams it down
	   everyone's throat
	Management listens too much to consultants
	Management wants to change e-business priorities
	Marketing defines new e-business initiatives with management
	   support	

Technical	Legacy systems support is too resource intensive There is a lack of available training for staff The technology requires a learning curve that is too long There is a lack of experience and knowledge of the technology The technology does not work Ordering and delivery of the technology are delayed A new version of the technology will be available soon There is a lack of support for interfaces to existing technology There is a gap in the technology The pilot results from the technology were not successful The performance of the technology is not satisfactory Features are missing The technology in use is obsolete The technology is not scalable to handle the workload The wrong direction in technology was taken	
Vendor/ Consultant	There is a lack of support from the vendor The vendor attempts to take over the project The vendor delivers something different from what was promised The vendor provides the wrong staff The vendor takes a different business direction, leaving you adrift The vendor's staff members are absent from the project too much The vendor's work is of poor quality The vendor's skills are insufficient, producing a gap The vendor's staff members seem to remain on the project forever There is no committed schedule from the vendor There is substantial turnover of vendor staff assigned to the project	

Area	Issue	Does issue apply?
Vendor/ **Consultant** (*cont'd*)	There are delays in vendor responses to problems, affecting the project The vendor's staff members are locked into their own methods, which are incompatible with the project The project is overdependent on the vendor Vendors don't cooperate among themselves Vendor subcontracts out the work and then disappears Vendor uses what they did on your project with their next client Vendor turns over work products that are not usable	

Index

The letters "ff" indicate the reader should also see the following pages.